BRITAIN'S INLAND ROAD BRIDGES

MARK CHATTERTON

This book is dedicated to my friend John Webb, who travelled all over the country with me in search of many of the bridges found in this book

Front cover: Bow Bridge Road Bridge in Chelmsford, Essex.

Back cover: Lower Kings Road Bridge in Berkhamsted, Hertfordshire.

First published 2025

Amberley Publishing
The Hill, Stroud,
Gloucestershire, GL5 4EP

www.amberley-books.com

Copyright © Mark Chatterton, 2025

The right of Mark Chatterton to be identified as the Author of this work has been asserted in accordance with the Copyright, Designs and Patents Act 1988.

All rights reserved. No part of this book may be reprinted or reproduced or utilised in any form or by any electronic, mechanical or other means, now known or hereafter invented, including photocopying and recording, or in any information storage or retrieval system, without the permission in writing from the Publishers.

ISBN: 978 1 3981 2643 5 (print)
ISBN: 978 1 3981 2644 2 (ebook)

British Library Cataloguing in Publication Data.
A catalogue record for this book is available from the British Library.

Typeset in 10pt on 13pt Celeste.
Typesetting by SJmagic DESIGN SERVICES, India.
Printed in UK.

Appointed GPSR EU Representative: Easy Access System Europe Oü, 16879218
Address: Mustamäe tee 50, 10621, Tallinn, Estonia
Contact Details: gpsr.requests@easproject.com, +358 40 500 3575

Contents

	Introduction	4
1	Southern and Western England	6
2	Eastern England	23
3	Central England	36
4	Northern England	58
5	Scotland	78
6	Wales	87
	Other Information	95
	Acknowledgements	96

Introduction

Britain's Inland Road Bridges is the follow-up book and companion volume to my previous book on bridges, *Britain's Coastal Road Bridges*. Once again it looks at 100 different road bridges, but this time ones that are found within Great Britain, away from the coast. As with the previous book, the road bridges covered should be still in use by motor vehicles, even if in one or two cases the bridge may be closed to road vehicles during the day.

The main problem with putting this book together was the wide choice of road bridges to be found all over Britain. This has meant that I have had to be quite selective about which to include. Again, I have tried to feature as wide a variety of bridges as possible. Apart from the common stone-arch type of bridge, I have included suspension, cable-stayed, arch, bascule, drawbridge and lattice bridges. Also, the bridges in these pages have some sort of historical, architectural, strategic or other such significance. Obviously the bridges that you find in the interior of Britain are not going to be as large or as long as those found on the coast. This is mainly because the rivers that they cross are much narrower than the various river estuaries that you find on the coast.

I have divided the book up into six different regions of Great Britain. Apart from Wales and Scotland, England has four different regions. These are Eastern England, Southern and Western England, Central England and Northern England. The Eastern England area includes bridges found mainly to the east of a line bordered by the A1 trunk road northwards from London. Southern and Western England covers the area to the south of a line from London to Bristol, roughly following the M4. The Northern England area is from a line starting at Chester in the west and going across the country via Sheffield to Hull. Finally, the Central England area is all the land remaining between the other areas.

Virtually all the big inland cities have at least one bridge featured. In some cases, I have included two different bridges in some of the larger cities like Manchester, Birmingham and Glasgow.

By far the majority of the road bridges in this book go over a river. However, there are some exceptions to this including bridges over canals, bridges over railways and stations, and even a bridge in a reservoir. There are also one or two viaducts, for example the Bromyard Viaduct and the Thelwall Viaduct.

Once again, during my research I couldn't help but be amazed at all the work that has gone into the building of these bridges, from the initial designers and planners with a vison for building a bridge in a particular place, through to the constructors, painters, welders and others who have created these bridges. I was also struck by the amount of old stone bridges which were damaged or even washed away by floods and storms in the past. Yet, human resilience has meant they were rebuilt, despite the setbacks. So, I take my hat off to all these people who have helped to provide us with the bridges that you see in this book.

One other thing that I noticed in my research was that interest in bridges has grown considerably in recent years with many videos of bridges being posted on various internet sites such as YouTube. On social media platforms like Facebook there are several groups which are dedicated to bridges in their various forms including *Bridge of the Day*, which has a massive 35,000 followers.

This isn't quite the end of my books on bridges though. The eagle-eyed among you reading this book will have noticed that I haven't included any bridges that are found in London. This is because I plan to write one more book on bridges which will look at the various bridges found in the capital – not just road bridges, but also railway and footbridges as well. So, enjoy the bridges in this book and marvel at their beauty and the engineering that went into building them.

1
Southern and Western England

Bickleigh Bridge, near Tiverton, Devon

Bickleigh Bridge carries the A396 road over the River Exe between Exeter and Tiverton in Devon. There was a bridge crossing here from at least the 1500s, though it was probably a packhorse bridge until 1809, when it was rebuilt after being badly damaged

Bickleigh Bridge in Devon. Some say this is the bridge that inspired Paul Simon to write the song 'Bridge Over Troubled Water'. (Photo © Lewis Clarke, CC by-SA 2.0)

by severe floods. The bridge that now stands was widened on its upstream side to 14 feet (4.3 metres) and was designed by Hiram Arthur. It has five semicircular arches made from Tiverton stone, with pointed cutwaters topped with pyramidal caps on both sides. Today the bridge is too narrow for two lanes of modern-day traffic, so traffic lights control the flow of traffic.

The river drops suddenly by the bridge and is often fast-flowing after heavy rainfall. On its south side is the Fisherman's Cot, a restaurant and hotel. The singer-songwriter Paul Simon is said to have stayed here in 1965 after playing a gig in nearby Exeter. This has led to some people speculating that Bickleigh Bridge was the inspiration for his song 'Bridge Over Troubled Water' as performed by Simon and Garfunkel. The story goes that on seeing the bridge and fast-flowing water from inside the hotel he was inspired to write the song.

City Bridge, Winchester, Hampshire

The City Bridge in Winchester, Hampshire, dates from the early 1800s and is situated at the junction of High Street and Bridge Street. It is small bridge of one arch and is 28 feet (8.5 metres) wide. It was designed by George Forder and was opened in 1813. It is constructed of limestone ashlar masonry in segmental stone with a stone balustrade along the top. It crosses the narrow River Itchen as it flows south towards Southampton. Previously there was a bridge nearby from the medieval period consisting of two arches, and before that a bridge of five spans from 852. It is sometimes known as Soke Bridge or St Swithin Bridge and is Grade I listed.

City Bridge in Winchester crosses the River Itchen and dates from 1813.

Clifton Suspension Bridge, Clifton, Bristol/Leigh Woods, Somerset

The Clifton Suspension Bridge over the gorge of the River Avon in Bristol is one of the most photographed bridges in Britain, along with Tower Bridge in London. It is also a popular tourist destination complete with its own visitor centre. It was designed by Isambard Kingdom Brunel at the age of just twenty-three. He became involved in the project after entering a competition to find the best bridge design over the Avon Gorge. However, it would take another thirty-three years for the bridge to come to fruition, by which time Brunel would be dead. It connects the Bristol district of Clifton on the north side with Leigh Woods in North Somerset on its south side and is also known simply as Clifton Bridge.

The origins of the bridge date back to 1754 when William Vick, a Bristol wine merchant, left £1,000 in his will to pay for a toll-free stone bridge to cross the River Avon. It was not until 1829 that things started happening when a competition to find a designer of the bridge was established. This did not yield any suitable candidates, so a second competition was launched. Brunel was the winner of this competition, and as a result his career as railway builder commenced. His original plans for the suspension bridge involved two Egyptian-style columns at either end which would each be topped with an Egyptian sphinx. The story of the bridge being built would be shown along the side of the bridge. However, it was soon realised that the columns would not take the weight of the sphinxes and so compromises were made.

The iconic Clifton Suspension Bridge in Bristol is the most well-known bridge in this book and one of the most photographed bridges in the world.

Construction work began in 1836 and stopped in 1842 when funds ran out. Brunel died in 1859, aged fifty-three, when work on the bridge had been halted. In 1860 construction began again after the Institution of Civil Engineers pledged to pay for the remaining costs. Sir John Hawkshaw and William Henry Barlow were now supervising the project. Barlow made some changes to Brunel's design including suggesting that the roadway be widened and adding a third set of suspension chains to the original two. The bridge was finally completed in summer 1864.

The wrought-iron chains that hold the bridge up are from the Hungerford suspension bridge, which had been demolished in 1860. The bridge weighs 1,500 tonnes and has a span of 702 feet (214 metres). It is 245 feet (75 metres) above the water below when it is high tide. It is a Grade I listed building. The bridge was originally designed for horse-drawn traffic and is only wide enough for single-file traffic, being 31 feet (9.4 metres) wide. Traffic flow is governed by traffic lights at either end of the bridge and it has a weight limit of 4 tonnes. There is a visitor centre on the south side which tells the story of the bridge. It is estimated that over 4 million vehicles cross the bridge every year.

Crane Bridge, Salisbury, Wiltshire

The Crane Bridge over the River Avon in central Salisbury, Wiltshire, is a medieval bridge dating originally from the fifteenth century, though another bridge from at least 1300 was on the same site before this one. It is a stone arch bridge consisting of four small, splayed

Crane Bridge is one of the smallest bridges to appear in this book and crosses the River Avon in Salisbury.

arches with projecting keystones and four cut waters. It was widened in 1898 when the south side of the bridge was taken down and then rebuilt for this purpose. It is situated just south of Fisherton Bridge further upstream, which used to be known as Upper Bridge, whilst Crane Bridge was known as Lower Bridge. Its present name comes from a nearby building called 'The Crane' and was not used until the sixteenth century. It is a Grade I listed building and joins up with Crane Street to the east.

Great Bridge, Tonbridge, Kent

Great Bridge (or Big Bridge) crosses over the River Medway at the north end of Tonbridge High Street in Kent, next to Tonbridge Castle. It has these names to distinguish it from a smaller bridge nearby which goes across Botany Stream, a tributary of the River Medway. Both bridges were built at the same time in 1887 by a local foundry which usually made cutlery, so locals have given them the nicknames of 'knife and fork' bridges. The main bridge replaced a stone bridge from 1776 and was made of cast iron. Half of the money for the bridge was given by The Rochester Bridge Trust, an organisation that helps with the upkeep of several bridges across the River Medway in Kent. The Trust again helped to pay for the cost of widening the bridge both in 1913 and in 1928. The bridge has pretty white latticed iron railings on either side of the roadway, along with six gas lamps.

Great Bridge (or Big Bridge) is a decorative bridge which crosses the River Medway in Tonbridge in Kent.

The bridge across the River Thames at Henley-on-Thames was the finishing point for the first ever Oxford and Cambridge Boat Race in 1829.

Henley Bridge, Henley-on-Thames, Oxfordshire/Berkshire

The road bridge at Henley-on-Thames was opened in 1786 and carries the A4130 across the River Thames to the east of Henley-on-Thames town centre. It consists of five semi-elliptical stone arches made from sandstone, with the central arch having a span of 40 feet (12 metres). It is 14 feet (4.3 metres) above the water at its central point. The bridge was designed by William Haywood of Shrewsbury who died in 1782 before work on the bridge had started. It was built by John Townsend of Oxford at a cost of around £10,000. It is Grade I listed and is sometimes called 'The Bridge of Sighs' by locals after the famous Venetian bridge.

There are sculptures of the Egyptian gods Isis and Tamesis, designed by Anne Seymour Damer, at the keystone of the central arch on each side of the bridge. The bridge was the finishing point for the first ever Oxford and Cambridge university boat race in June 1829. In 2011 the bridge underwent £200,000 of repairs after it had been damaged by a boat hitting it.

Maidenhead Bridge, Maidenhead, Berkshire/Taplow, Buckinghamshire

Maidenhead Bridge carries the A4 trunk road across the River Thames to the east of Maidenhead in Berkshire and joins with the town to Taplow, in Buckinghamshire. It is

Maidenhead Bridge dates from 1777 and was a toll bridge until 1 November 1903.

Grade I listed with the current bridge here dating from 1777. It replaced a previous bridge here from 1280 which was deemed to be beyond repair. The replacement bridge had thirteen Portland stone arches and was designed by Robert Taylor with an estimated cost of £25,000. As this was thought to be too expensive only the arches over the river were made of Pentland stone, with the rest being built from brick. John Townsend of Oxford was the builder, with the total cost coming in at £19,000 (approximately £3 million in today's money). At first it was tolled, but the toll charges were eventually abolished on 1 November 1903. On this day a large crowd took the toll gates and threw them into the River Thames!

Maidstone Bridge, Maidstone, Kent

Maidstone Bridge over the River Medway in Kent was opened in 1879, replacing a medieval bridge which had been on the same site since the fourteenth century. It links the centre of Maidstone with its southern suburbs. The current bridge was engineered by John Bazalgette, along with its approaches, at a cost of £48,000, with a large portion of the costs being met by the Rochester Bridge Trust. Bazalgette used a three-arch design, consisting of large Cornish granite blocks as opposed to a single-span iron construction, as put forward by others. Work on the bridge started in October 1877 and lasted nearly two years. It was

Maidstone Bridge, which opened in 1879, crosses the River Medway in Kent.

formally opened on 6 August 1879 by the mayor of Maidstone, Charles Ellis. The final cost was in the region of £50,000.

The centre arch is 54 feet (16.5 metres) across and the adjacent arches are each 47.5 feet (14.5 metres) across. The roadway width is 24 feet (7.3 metres) with adjacent 8-foot-wide (2.4 metres) footpaths. The bridge was widened between 1926 and 1936. It was built within inches to the north of the old bridge. The original bridge had seven arches at first and was wide enough for only one vehicle. It was subsequently widened for two carriages to pass and in 1858 a further 8 feet of width was added to it. This bridge was removed after the opening of the new bridge in 1879.

Marlow Bridge, Marlow, Buckinghamshire/Bisham, Berkshire

Marlow Bridge is a suspension bridge which crosses the River Thames between the town of Marlow in Buckinghamshire and the village of Bisham in Berkshire. It was opened in 1832 after three years of construction. It replaced a nearby timber bridge dating from 1789 which fell into the river in 1828. It was designed by William Tierney Clark, who also designed the similar-looking and larger Széchenyi Chain Bridge across the River Danube in Budapest, Hungary. It is a Grade I listed building and now has a 3-tonne weight limit on it. It underwent restoration work in 1965 and from 1972 onwards through traffic started to use the newly constructed Marlow Bypass Bridge instead. However, in September 2016 a 37-tonne heavy goods lorry from Lithuania tried to drive across the bridge, forcing the bridge to be closed for two months whilst structural tests were carried out. No serious damage was reported, though much of the exposed steelwork needed repainting.

Marlow Bridge was designed by William Tierney Clark, who later designed the larger and similar-looking Széchenyi Chain Bridge across the River Danube in Budapest, Hungary.

Prince Street Bridge, Bristol

The Prince Street Bridge in Bristol is an iron swing bridge which crosses Bristol Harbour. It was opened in 1879 and replaced a previous bridge which had been here since 1809. It uses hydraulic power to make it open which comes from a nearby engine house and accumulator tower. Originally the bridge carried two-way traffic across it, but now it is one-way only southbound to the M Shed in Bristol Docks, whilst the other lane is for

Prince Street Bridge by Bristol Docks is used by many more pedestrians than motorists and may well be pedestrianised in the future.

pedestrians and cyclists. A recent survey showed that fewer than 2,000 cars a day use the bridge, whilst 24,000 pedestrians and cyclists use it. This has led to a proposal for the bridge to become pedestrian only in the future. The bridge was closed between 2015 and 2017 for repair work to the hydraulic system.

Pulteney Bridge, Bath, Somerset

The Pulteney Bridge carries Argyle Street over the River Avon in Bath and is one of just four bridges in the world to have shops built onto it. Its name comes from William Johnstone Pulteney, a local landowner who wanted a bridge to match his aspirations of building a new town, Bathwick, on the east side of Bath, as well as being a replacement for the ferry here. The bridge was dedicated to, and named after, his wife, Frances Pulteney.

The bridge was designed by Robert Adam in 1769 and is said to have been inspired by Florence's Ponte Vecchio bridge in Italy. It was opened in 1774 at a cost of £10,000. It reflects the Palladian architecture of the Georgian period with pilasters, pediments and lead domes. It was one of the last bridges to be built in Britain that had houses and shops built into it. However, in 1800 the bridge collapsed on its northern side, taking several buildings on the bridge into the River Avon with it. On reconstruction of the bridge a few

The world-famous Pulteney Bridge in Bath is just one of a few bridges in the world that has shops built onto it.

years later, the rebuilt shops looked different to those that survived the collapse and this can be noticed as you walk across the bridge.

The bridge is 58 feet (17.7 metres) wide and it has three segmental arches which are 33 feet (10 metres) across. It is right next to the crescent weir, though it is not possible to see the river below from the bridge itself. Today, only buses, taxis and delivery vehicles are permitted to cross the bridge. It was featured in the 2012 film *Les Misérables* where Javert throws himself into the River Seine, with Bath doubling up as Paris.

Reading Bridge, Reading, Berkshire

Reading Bridge carries the B3345 across the River Thames to the north of Reading station and the town centre. It consists of one main span across the water with two smaller arches at each end over the footpaths by the side of the river. The bridge is made from reinforced concrete and when it was opened in October 1923 was the longest of its type in Britain. Originally it was planned to use steel as the bridge material, but this was changed to reinforced concrete in 1913 when this new material started to be used in building projects and was deemed to be more longer-lasting and cheaper to maintain. The outbreak of the First World War in 1914 delayed the bridge's construction, which didn't commence until 1922. It was designed by L. G. Mouchel & Partners with Holloway Brothers undertaking construction work. The main span is 180 feet (55 metres) across with a clearance of 18 feet (5.5 metres) above the water. The bridge deck is 40 feet (12 metres) wide with a 27-foot-wide (8.2 metres) carriageway.

Reading Bridge across the River Thames in central Reading was the longest concrete bridge in Britain when it was opened in 1923.

Staines Bridge, Staines-upon-Thames, Surrey

Staines Bridge crosses the River Thames at Staines-upon-Thames in Surrey. It originally carried the A30 trunk road, but this number was given to the Staines bypass, so the road is now the A308. There has been a bridge at this site since Roman times and this is the fourth bridge to be built here. The current bridge was built in the late 1820s, and was opened in April 1832 by King William IV. It has three arches above the river, with four more on the north side and two on the south side. It is constructed of white granite blocks sourced from Aberdeen. It has a clearance of 19.5 feet (6 metres) above the river. Tolls were applied at first, but these were stopped in 1871. In 1956 pavements were added. It was designed by George Rennie and is similar in design to Waterloo Bridge. It is Grade II listed and has a 7.5-tonne weight limit.

Opened in 1823, Staines Bridge across the River Thames was designed by George Rennie, whose father had designed the original Waterloo Bridge in London.

Tone or Taunton Bridge crosses over the River Tone in Taunton, Somerset.

Tone Bridge, Taunton, Somerset

Tone Bridge or Taunton Bridge carries the A3027 road across the River Tone in Taunton, Somerset. It consists of a series of cast-iron girders supported by two rows of circular columns and weatherings. On the bridge are four double lamp standards which are believed to be one of the earliest examples of electric street lighting in a town in Britain, dating from when the bridge was built. The bridge dates from 1894–95 and replaced a previous bridge which had stood here since medieval times. It was designed by J. H. Smith, the Borough Surveyor, and cost £7,000. In 1936–38 the girders and deck were replaced with steel ones due to wear and tear, but the cast-iron parapets and lamp standards were left in their original form. It is often referred to as the River Tone Bridge by locals.

Town Bridge, Arundel, West Sussex

Town Bridge crosses the River Arun in Arundel, West Sussex. It connects the main part of the town with its outskirts to the south-east. The bridge is believed to have first opened in the 1100s and was the limit of Arundel port, which flourished in the Middle Ages right through until the late 1800s. It was the lowest bridge crossing of the river until 1908 when the Littlehampton swing bridge was opened several miles further south. The current three-arch stone bridge was widened in 1934, and its steep approach roads were changed to a more acceptable level, as they were deemed too dangerous for the ever-increasing amounts of motor traffic using the bridge. The bridge, also known as Arundel Bridge, was the main crossing point of the river in the region until a relief road was opened in 1973.

Town Bridge in Arundel, West Sussex, was for many years the nearest bridge crossing of the River Arun to the sea until 1908 when a bridge was opened at Littlehampton.

During the Second World War the town of Arundel was seen as being on the front line of a possible German invasion along the south coast. As a result of this threat, the bridge had forty deep concrete sockets built into it which held various lengths of temporary railway track. The bridge would be destroyed as a final form of defence if necessary and had explosives attached to it, which would only be used as a last resort.

Town Bridge, Bradford-on-Avon, Somerset

Town Bridge crosses the River Avon in Bradford-on-Avon, Somerset, and replaced a ford ('Broad Ford') that gave Bradford its name. A packhorse bridge, dating from medieval times, is believed to have stood here previously. Certainly, the bridge is mentioned as far back as 1200, and it suffered from flood damage around 1340. It was quite narrow for many years, but was widened on the western side in 1769. It has nine arches built of Cotswold stone. The two arches on the southern side are the oldest, dating from medieval times, and are more pointed than the rest of the bridge.

On the southern end of the bridge is an old lock-up, built on the foundations of a former chapel. Drunks and troublemakers would be locked up here overnight. Both the bridge and lock-up are Grade I listed.

Parts of the nine-arch bridge at Bradford-on Avon, Somerset, date back to medieval times.

Town Bridge, Trowbridge, Wiltshire

Trowbridge is a small town of 37,000 in Wiltshire around 8 miles (13 km) south-east of Bath. It grew due to the wool trade in the Middle Ages and was also known as a brewing centre. The name is thought to have come from the words 'Tree Bridge', referring to the earliest bridge crossing of the River Biss which runs through the town. An alternative theory is that the name comes from 'the bridge by Trowle', being the name of a nearby hamlet. The current bridge is known simply as 'Town Bridge' and goes across the River Biss in the centre of the town. It dates from 1777, as shown in a date stone above the centre of its three segmental arches. This states that the bridge was built by Esau Reynolds, a local architect. In 1757 the Blind House was built by the bridge as a lock-up for drunkards and criminals until the police station was built in 1854. The bridge carries the road called Wicker Hill between the railway station to the south and the town centre to the east.

The Georgian Town Bridge at Trowbridge in Wiltshire dates from 1777.

Walton Bridge, Walton-on-Thames, Surrey

Walton Bridge carries the A244 across the River Thames in Surrey linking Walton-on-Thames to the south with Shepperton in the north. It is the sixth bridge to have been erected on this site and was opened on 22 July 2013, six months ahead of schedule. The previous five bridges dated from 1750, 1788, 1864, 1953 and 1999, with the last two being considered as

The current Walton Bridge at Walton-on-Thames in Surrey is the sixth different bridge to be built here and is known as a thrust arch bridge.

temporary bridges. It is a thrust arch bridge with two parallel arches made from steel and it is just over 18 feet (5.5 metres) high. Although the planning stages for the bridge started in June 2005, it wasn't until January 2011 that full planning permission was granted. Costain won the £32 million contract to build the bridge and Atkins were the designers. Originally a box girder design was considered, but this would have meant steeper gradients to and from the bridge, so an arch was chosen as the best alternative. The bridge won the best practice award at the 2014 British Construction Industry Awards.

Westgate Bridge, Canterbury, Kent

This nineteenth-century bridge crosses the River Great Stour as it flows around the north side of Canterbury town centre. It gets its name from the nearby Westgate Gatehouse which is part of the walls of the city and is the largest surviving city gate in Britain. It carries St Dunstan's Street from the north into the city centre on two lanes of traffic, with southbound traffic going through the gate and northbound traffic around it. Originally there was a medieval bridge here, which was replaced by a new bridge in Tudor times. This lasted until 1828 when the present bridge was built. It was widened to cope with two lanes of traffic sometime in the early twentieth century. It is approximately 33 feet (10 metres) wide and the arch is 6 feet (2 metres) above the usual water level.

The small Westgate Bridge in Canterbury, Kent, gets its name from the imposing Westgate next to the bridge.

2
Eastern England

Ballingdon Bridge, Sudbury, Suffolk/Ballingdon, Essex

Ballingdon Bridge between Sudbury in Suffolk and Ballingdon in Essex carries the A131 Braintree Road across the River Stour. It marks the border between Suffolk and Essex and is on a bridge site which has existed since Roman times. The current bridge was built between 2002 and 2003 and was opened on 18 July 2003. The bridge design was chosen

Ballingdon Bridge at Sudbury in Suffolk is mentioned in the children's book *101 Dalmatians*, written by local author Dodie Smith.

from a shortlist of five by the people of Sudbury. It was designed by the firm of Brookes, Stacey and Randell and was constructed by Ove Arup using pre-cast concrete sections, with an overall length of 112 feet (34 metres). Aluminium was used in the balustrade and the illuminated bollards, which were designed to avoid the need for lamp posts on the bridge. It replaced a previous bridge from 1911 which could no longer cope with the heavy lorries using the bridge. It was designed with a 120-year lifespan. The bridge is mentioned in the children's story *101 Dalmatians* by Dodie Smith, who lived in the area. The two parent dogs, Pongo and Perdita, are described as crossing the bridge on their way into Sudbury in search of their stolen puppies. The bridge has won several awards for its unique design and lighting.

Battlesbridge Bridge, Battlesbridge, Essex

The present bridge at Battlesbridge was opened in 1872, several bridges having existed here previously, the earliest dating back to 1372. It holds a single-track road and was widened to two lanes in the 1970s. It was originally numbered as the A130, the number of the main road between Canvey Island and Chelmsford. When a bypass was built further to the west in the 1990s, it became the A1245. It has a span of 119 feet (36.4 metres) with two piers, and there is a headroom of 10 feet (3 metres) to cope with high tides that come up the River Crouch. Battlesbridge has been a small port since medieval times, reaching its peak in the late nineteenth century when cargoes of coal, malt, lime and chalk passed through the port. The name is not thought to have come from a battle here, but rather from a local family called 'Bataile'. Battlesbridge is now well known as an antiques centre.

The bridge over the River Crouch at Battlesbridge in Essex at low tide.

Bow Bridge Road Bridge in Chelmsford, Essex, is the newest bridge to be featured in this book, being opened in 2025.

Bow Bridge Road Bridge, Chelmsford, Essex

This is the newest bridge in this book and opened on 11 July 2025. It crosses the River Chelmer in Chelmsford, Essex. It is a bowstring tied arch bridge made of Corten steel, with a span of 161 feet (49 metres) and a width of 20 feet (6 metres), which is wide enough for two lanes of traffic. It was constructed by Graham and designed by Arcadis. It was built to provide access to a new development on the north side of the River Chelmer called Chelmer Waterside, connecting Parkway in the south with Wharf Road in the north. Previously the city's gasworks used the site of the new housing development. It has a speed limit of 20 mph and is high enough above the water below to allow small boats to pass underneath. Local residents were asked to name the bridge and out of five choices they voted for 'Bow Bridge Road'. The other names were John Rennie Bridge Road, Timespan Bridge Road, Confluence Bridge Approach and Jabez Church Bridge Approach.

Carrow Bridge, Norwich, Norfolk

Carrow Bridge takes the A147 Carrow Road across the River Wensum to the south of the centre of Norwich in Norfolk. The present Carrow Bridge was opened in June 1923 by the Prince of Wales, the future King Edward VIII. It cost £42,000 and replaced two previous

Carrow Bridge in Norwich was opened in 1923 as a bascule lifting bridge for river traffic travelling along the River Wensum.

bridges that were situated 500 yards further upstream. This was partly due to pressure from Colmans, the mustard manufacturer, whose factory was expanding on both sides of the river and needed a bridge for ease of access to their factories. It was also planned to give much-needed work to the unemployed of Norwich at that time. The planned crossing site for the new bridge meant that Carrow Road had to be diverted to fit in with the new Carrow Bridge. It is a single-leaf roller bascule lifting bridge and was designed by A. E. Collins, the city engineer for Norfolk. It was built by J. Butler and Co. of Leeds and was often raised for the busy river traffic going along the River Wensum. However, as the bridge is hardly lifted nowadays there have been suggestions to weld the bridge shut.

Crescent Bridge, Peterborough, Cambridgeshire

Crescent Bridge is a steel bowstring lattice bridge which goes over the East Coast mainline railway to the south of Peterborough railway station. It was opened in 1913 and was built to replace two level crossings which caused long delays for traffic and pedestrians going over the railway line whenever a train was going through. At present ten different railway lines pass under the bridge. Back when it was built there were many more, so there was a strong need for a bridge. The bridge cost £34,000 to build and was paid for by the Great Northern Railway. It was opened on 16 April 1913 by the Mayoress of Peterborough, Mrs J. G. Bartford. The bridge joins Thorpe Road on its west side with the A15 to the east. The name comes from a terrace of Victorian houses in a street called The Crescent which

Crescent Bridge in Peterborough crosses over the East Coast mainline and is a well-known landmark in the city.

was demolished to make way for the bridge. The steel framework of the bridge was built by the Cleveland Bridge and Engineering Company of Darlington. It is painted in a distinct blue colour and is a well-known feature of the Peterborough skyline.

Foundry Bridge, Norwich, Norfolk

This single-span bridge crosses the River Wensum and carries the A1242 Prince of Wales Road east out of the city centre past Norwich railway station. The previous Foundry Bridge opened in 1844 at the same time as Norwich Thorpe station. It replaced an earlier wooden and

Foundry Bridge connects the railway station with the city centre in Norwich.

stone bridge which had opened in 1811 but was too low for some boats to pass underneath it. The cast-iron bridge was designed by C. Atkinson and gets its name from a nearby iron foundry. As time went on this bridge was replaced when Norwich Thorpe railway station was being rebuilt in 1886. The new bridge was opened in January 1888 at a cost of £12,000. It is a Grade II listed building and is 55 feet (16.8 metres) long and 50 feet (15.3 metres) wide. The bridge was apparently assembled in the station yard and then rolled across the road onto the existing stone abutments. The Great Eastern Hotel stood beside the bridge for many years until, in 1963, it was demolished. It was replaced by the Hotel Nelson, which opened in March 1971. This in turn was then replaced by a Premier Inn.

High Bridge, Lincoln, Lincolnshire

High Bridge in Lincoln carries the High Street over the River Witham and is the oldest bridge in Britain with buildings on it, dating from approximately 1160. The row of shops on the west side of the High Street date from Tudor times – hence their black and white design. The present tea and coffee shop has used the building since 1937. Since 1971, when a bypass was built around Lincoln, through traffic does not use this part of the High Street

High Bridge in the centre of Lincoln crosses over the River Witham and is the oldest bridge in Britain that still has buildings on it.

anymore. However, delivery vehicles and emergency vehicles are still able to access the High Street and cross the bridge here.

As the River Witham becomes very narrow as it goes under the High Street at this point, floods are a problem and have been since the fourteenth century. After heavy rain the raised water level means that boats cannot navigate under the bridge. This area has been called 'The Glory Hole' due to its similarity to the halos over saints and holy people. This replaced the previous nickname of 'The Murder Hole', as this was the place where dead bodies that had been thrown into the river frequently got stuck! A stone obelisk was placed on the bridge in 1762 as a conduit, replacing a chapel which had been there. This was eventually moved in 1939 due to concerns about its weight affecting the bridge and it blocking traffic using the bridge. The bridge is a Grade I listed building and also a scheduled monument.

Magdalene Bridge, Cambridge, Cambridgeshire

Magdalene or Great Bridge was opened in 1823 and takes Magdalene Street across the River Cam to the north of central Cambridge. It has a single cast-iron span and was designed by Arthur Browne and Benjamin Brown of Norwich. It cost £2,350 to build but had to be strengthened in 1832 to cope with increased traffic. There are ashlar piers at either end joined by iron railings. It is a Grade II listed building and in 1982 was strengthened with a steel frame after subsidence had weakened the bridge. Previously there were several wooden bridges here until 1754 when the first stone bridge was built. It is thought that Cambridge gets its name from an earlier bridge that crossed the River Cam here.

Magdalene Bridge in Cambridge is believed to be near to the crossing point of the River Cam where the name 'Cambridge' comes from.

Silver Street Bridge, Cambridge, Cambridgeshire

Silver Street Bridge (or Small Bridge) goes over the River Cam in Cambridge at the entrance of the Backs from the Mill Pond area by Laundress Green. This is where the River Cam splits into two parts south of the bridge with the main part of the river going on one side of Laundress Green and the smaller stream going south to the west of Sheep's Green. The current Grade II listed bridge is relatively recent, having been opened in 1959. It had been designed by Sir Edwin Lutyens in 1932 as a reinforced concrete bridge, which rested on rafts. However, the Royal Arts Commission said that the bridge had to be clad in Portland stone to help preserve the character of the area. It replaced a cast-iron bridge that had been built in 1841, which had been weakened by the floods of 1947. This in turn was built to replace the previous bridge here which had become 'dilapidated' according to a report made in 1836. If you stand on Silver Street Bridge looking north along the River Cam you get a view of the well-known Mathematical Bridge, as well as the constant stream of punts sailing along the river.

Silver Street Bridge in Cambridge is at the entrance to 'the Backs' where the River Cam passes the back part of several Cambridge University colleges and is popular for punting.

Town Bridge in Boston, Lincolnshire, dates from 1913 and was designed by John J. Webster.

Town Bridge, Boston, Lincolnshire

The current Town Bridge in Boston, Lincolnshire, is a single-span arch bridge over the River Witham made from steel and opened on 18 July 1913. It replaced a cast-iron bridge known locally as Rennie's Bridge (after the architect Sir John Rennie), which had been built on the same site between 1806 and 1808. There were several problems with this bridge, not least the fact that various cracks kept appearing in the cast iron when it was first built. This in turn had replaced various bridges which had stood here since the 1200s.

The 1913 bridge was designed by John J. Webster and used the sandstone abutments from the previous bridge. It is 43 feet (13 metres) wide, including a 25-foot (7.6 metre) carriageway, and has been Grade II listed since 1975. In 1964 a new bridge bypassing the town centre was built further downstream at The Haven, reducing traffic using the bridge. It is now one-way in a northerly direction, with a bus lane going the other way. In the summer of 2020, the bridge underwent blast cleaning and complete repainting by the company Jack Tighe Ltd due to corrosion caused by sea salt from tidal water passing under it.

Town Bridge, Peterborough, Cambridgeshire

The present Town Bridge over the River Nene in Peterborough dates from 1934. It was built to replace the previous Town Bridge which had been opened in 1872. This was constructed of cast iron and was meant to last much longer than it did, having replaced a wooden bridge which had stood here for over 500 years. Work on the present Town Bridge started

The bridge known as Town Bridge in Peterborough crosses the River Nene and dates from 1934.

in 1932 and concrete was chosen as it would be more resilient than the iron of the previous bridge, which was not demolished until the concrete bridge was completed. The new bridge was wide enough for four lanes of traffic and as a result the road called Narrow Street which led onto the bridge on its northern side was demolished to make way for a much wider roadway called Bridge Street, along with a new town hall. The bridge carries the A15 London Road over the river in a north–south direction.

Town Bridge, Stamford, Lincolnshire

Town Bridge, or Stamford Bridge, takes the A1175 across the River Welland in Stamford, Lincolnshire. It was constructed in 1849 and was designed by Edward Browning. It replaced a medieval bridge from the twelfth century. It connects Stamford town centre to the north

Town Bridge over the River Welland at Stamford in Lincolnshire was built in 1849 and replaced a medieval bridge which had previously been on this site.

with Stamford Baron to the south. Finance for the 1849 bridge came from the Midland Railway who wanted to build the railway across the previous bridge here, but this never materialised, mainly because the local landowner, the Marquess of Salisbury, opposed it. Eventually a compromise was reached with the railway company paying £5,000 for the bridge to be built and the Marquess paying the rest. The total cost was around £8,000.

The bridge consists of three segmental arches, each with a 30-foot (9.1 metre) span and a similar-sized carriageway. It was completed in March 1849, but it didn't open formally until 1 May when the toll house at the north end was finished. Tolls were not popular and these were finally ended in 1868. In 1880 floods nearly covered the bridge. They were the highest floods for 200 years, but the bridge survived. Town Bridge used to carry the A1 London to Edinburgh Road over the river until 1960 when a new dual carriageway was built on the west side of the town. In 2017 the four original lamps on the bridge were cleaned, restored and fitted with LED lighting. The bridge is subject to a 7.5-tonne weight limit for HGVs, and has three-way traffic lights to help with traffic flow. It is Grade II listed.

Trent Bridge, Newark-on-Trent, Nottinghamshire

Trent Bridge at Newark-on-Trent in Nottinghamshire is a stone and brick bridge of seven arches which dates from 1775. It carries the Great North Road across the River Trent next to the ruins of Newark Castle. For many years this road was the A1 between London and Edinburgh, until the A1 bypass was built to the east of the town in the 1990s. It replaced a twelfth-century wooden bridge which had been built around 1135. However, the number of

Newark Bridge crossing the River Trent in the shadow of Newark Castle.

floods here meant that the bridge was often damaged and needed repairing. It is 170 feet (52 metres) in length and 38 feet (11.5 metres) wide and was built under the supervision of Stephen Wright. It is essentially constructed of bricks with stone facades. In 1848–49 railings and footways were added due to the state of the roadway. The bridge was also widened to cope with the increased number of people using the recently opened Newark Castle railway station.

Wansford Bridge, Wansford, Cambridgeshire

The bridge at Wansford, around 10 miles west of Peterborough, used to carry the A1 Great North Road over the River Nene here. Being only wide enough to carry single-file traffic it was a bottleneck for many years, so in 1929 a new bridge over the River Nene was opened a few hundred yards to the east. Wansford Bridge consists of a series of twelve arches of varying sizes and was believed to have been built sometime in the 1200s, with alterations made at different times. We do know that a storm damaged the bridge in 1571, with the bridge being rebuilt in 1577. Three arches were added to the north end in 1672 and two more were added on the southern side in 1795 after ice damage. These latter two cross the river itself, with the other arches being

The old bridge at Wansford in Cambridgeshire dates from the 1200s and used to carry the Great North Road across the River Nene here.

used to cross the river's flood plain over a meadow. The bridge being quite narrow, it has several pedestrian refuges on both sides. After the A1 was moved off the bridge, it was numbered the A6118 for many years but is now numbered the C340. The roadway over the bridge is called Bridge End.

Wroxham Bridge, Hoverton/Wroxham, Norfolk

Wroxham Bridge lies in the heart of the Norfolk Broads and takes the A1151 Norwich Road over the River Bure. It connects Wroxham on the south side with Hoverton on the north side of the bridge. Originally there was a wooden bridge built here in 1569. This was rebuilt in 1619 out of brick and stone as a single-arch bridge. Elements of this bridge can still be seen on the north-west side, whilst the south-east side of the bridge shows signs of having been widened, probably in the eighteenth or nineteenth centuries. In 1969 a Bailey bridge was placed over the bridge, then in 1992 a steel-girdered deck or 'umbrella bridge' was added on top of this. It is thought to be a difficult bridge to navigate under by boat and because of this there is a pilot station on the Hoverton side of the river where vessels can arrange to be guided under the bridge. It is a scheduled monument.

Wroxham Bridge in Norfolk, in the heart of the Norfolk Broads, crosses over the River Bure and parts of it date from the 1600s.

3

Central England

Abingdon Bridge, Abingdon, Oxfordshire

Abingdon is a small town around 10 miles south of Oxford which the River Thames passes through. Abingdon Bridge takes the A415 across the River Thames in Oxfordshire and was built between 1416 and 1422. The bridge is made up of two parts: the main Abingdon Bridge, which has six arches and is on the town side of the river; and Burford Bridge, which has one main arch over the river with a further four arches over the water and two more over the flood plain. They were added to the bridge in 1453. They are joined by Nag's Head Island. It is made from local limestone.

The bridge has been altered and widened at various times. In 1790 one of the arches was widened and raised to help with navigation. More widening took place in 1829–30. Then in 1927 three arches from the fifteenth century were knocked down to make one wider arch to help with navigation.

The bridge across the River Thames at Abingdon in Oxfordshire dates from the 1400s.

Bedford Bridge crosses the River Ouse and was opened in 1811, replacing a previous bridge from medieval times.

Bedford Bridge, Bedford, Bedfordshire

Bedford Bridge, or Town Bridge as it is also known, was opened in 1813 and crosses the River Ouse in central Bedford. It replaced a previous bridge that had been on the same site since the 1100s. This bridge had two gatehouses and was demolished in 1811 to make way for the present Georgian bridge. It was designed by John Wing and was formally opened on 1 November 1813. It cost around £15,000 and tolls were charged until 1835 to help pay for its costs. It has five arches and is built from local sandstone. From 1938 to 1940 the bridge was widened on its western side using concrete to cope with the increase in motor vehicles using it. In the 1980s remedial work on the bridge's stonework took place.

Black Sabbath Bridge, Birmingham, West Midlands

Unlike most big cities in Britain, Birmingham is not situated next to a major river. However, due to its industrial past there are over 100 miles of canals with many bridges crossing them. One such bridge in central Birmingham is the Broad Street Bridge which takes Broad Street over the Birmingham Canal Old Line to Gas Street Basin where it connects to the Worcester and Birmingham Canal. It is thought that the canal was here before the bridge, which was built around 1778. It was then widened as road traffic increased. The bridge is also known as the Broad Street Tunnel at canal level and is 98 feet (30 metres) long.

In 2019 it was renamed 'Black Sabbath Bridge' in honour of the group of that name who achieved worldwide fame in the 1970s, becoming Birmingham's most famous rock group.

The Black Sabbath Bridge in Birmingham is a popular tourist destination with a dedicated bench and images of the four original members of the rock group on Broad Street.

There is now a bench on the north side of the bridge featuring images of the original four members of the band, which has become a popular tourist attraction. There is also a large white cross bearing the legend 'Black Sabbath Bridge' across it. It was officially renamed on 26 June 2019 when Tony Iommi and Geezer Butler of Black Sabbath arrived by narrowboat on the canal beneath it to perform the opening ceremony.

Bromford Viaduct, Birmingham, West Midlands

One of three viaducts included in this book, the Bromford Viaduct in Birmingham is included as it is regarded by many as the longest road bridge in Great Britain. It carries the M6 motorway between junctions 5 (Castle Bromwich) and 6 (Gravelly Hill) in Birmingham along the valley of the River Tame. It is 3.5 miles (5.5 km) long and was constructed between 1961 and 1972. It is held up by a series of giant concrete columns. Due to water erosion to the concrete at some of the bases of the columns, National Highways carried out work to repair this damage in 2022 at a cost of £1.5 million. They also relocated electrical cables here. The name 'Bromford' comes from the Old English 'brom ford' where pieces of broom wood were used to ford or cross the River Tame in this area.

The Bromford Viaduct which carries the M6 motorway along the valley of the River Tame in north Birmingham is said to be Britain's longest road bridge.

Clopton Bridge, Stratford-upon-Avon, Warwickshire

Clopton Bridge crosses the River Avon at Stratford-upon-Avon in Warwickshire at a point in the river where there was a ford in the past. This is where the name Stratford-upon-Avon is believed to have come from. The bridge dates from approximately 1484 and is thought to have been named after Hugh Clopton of Clopton House, who financed its cost. It consists of fourteen painted arches built from local stone, with more arches forming a causeway over the flood plain on its north side. It carries the A3400 road in two lanes into the eastern

Clopton Bridge in Stratford-upon-Avon dates from around 1484 and has fourteen arches crossing over the River Avon.

side of Stratford-upon-Avon. In 1811–12 the bridge was widened on its northern side and a few years later a ten-sided toll house was built at the town end of the bridge. Then in 1827 a cast-iron footbridge was added to the northern side. The bridge is Grade I listed and is also a scheduled monument.

English Bridge, Shrewsbury, Shropshire

The English Bridge in Shrewsbury, Shropshire, is one of two road bridges which cross the River Severn in the city. The other, known as the Welsh Bridge, is so called as it is the nearest of the two to the Welsh border. The English Bridge is a masonry seven-arch bridge which was opened on 26 October 1927 by Queen Mary and cost £86,000 to construct. It was designed by Arthur W. Ward, the Borough Surveyor for Shrewsbury, and replaced an earlier bridge from 1769 which had a much steeper central arch to allow greater headroom for boats passing underneath. Ward managed to reduce the height of this central arch by 5 feet (1.5 metres) and also widen the bridge to 50 feet (15 metres) compared with its original 23.5-foot (7 metre) width. Prior to the 1769 bridge, a bridge had stood here since Norman times which had five arches and contained several houses and shops built onto it. The A5 trunk road between London and Holyhead originally crossed the bridge, but this was diverted around the southern edge of Shrewsbury in the 1980s. It is now numbered the A5191. The bridge links the city centre with the Abbey Foregate area.

English Bridge in Shrewsbury is a seven-arch stone bridge across the River Severn and was opened in 1927.

Exeter Bridge in central Derby is named after Exeter House which once stood by the side of the River Derwent here.

Exeter Bridge, Derby, Derbyshire

Exeter Bridge is found to the east of Derby city centre and crosses the River Derwent. Its name comes from Exeter House, a large building on the south bank which was owned by the Bingham family. They had built a wooden footbridge to the other side where their gardens were situated. This bridge and the house were demolished in 1929 and the present bridge erected.

The current bridge was designed by Charles Herbert Aslin from the City of Derby Architect's Department. It is a single-span concrete structure. During its construction several heavy vehicles were driven across the bridge to see if it could cope with such heavy weights! The bridge withstood this test, though what would have happened if the bridge collapsed is unthinkable.

Exeter Bridge was opened on 13 March 1931 by Herbert Morrison, the Minister of Transport. The bridge features bas-relief sculptures of local men John Lombe, William Hutton, Herbert Spencer and Erasmus Darwin.

Folly Bridge, Oxford, Oxfordshire

Folly Bridge crosses Folly Island and two channels of the River Thames as it flows through Oxford. It carries the A4144 Abingdon Road south from Oxford city centre. It is believed that this area is the origin of the name 'Oxford' as this is where oxen were originally driven across a ford here. Certainly, there has been a bridge crossing here since Saxon times with a stone bridge being constructed by Robert d'Oilli in 1085, replacing

Folly Bridge in Oxford is believed to have been named after a castle-like structure that was built on top of one of the gatehouses on the bridge in 1611.

a wooden structure on this site. The stone bridge was then known as South Bridge and south of it was a long causeway called Grandpont, which crossed a large area of low-lying land which was prone to frequent flooding from the Thames, or Isis as it is also known. The current name of the bridge, 'Folly Bridge', is believed to have originated in the 1600s when a man called Thomas Welcome built an extra floor on top of a gatehouse on the bridge in 1611. The new addition had castle-like features and mullioned windows, leading to some people calling it 'Welcome's Folly'.

However, the present-day bridge dates from 1827 when it was decided that the previous bridge was in need of urgent repairs when its foundations had become unsafe, the weir underneath the bridge contributing to the problem. It took two years to build and was designed by Ebenezer Perry of London. It is built of Cotswold sandstone with three arches connecting it to Folly Island with a further arch over another channel of the Thames. Tolls to cross the bridge were charged for many years until they were abolished in 1850. A toll house on the north end of the bridge was rebuilt in 1844. This and the bridge itself are Grade I listed.

Jackfield Bridge, near Ironbridge, Shropshire

Jackfield Bridge (or Jackfield Free Bridge as it is also known) crosses the River Severn half a mile downstream from the world-famous Iron Bridge in Shropshire. It replaced the former Free Bridge which had been here since 1909. Its name comes from the fact that tolls were not charged to cross the river here, unlike the nearby Iron and Coalport bridges. This bridge deteriorated over the years and by 1993 it was deemed to be beyond economic repair. It was then demolished despite the fact that it was Grade II listed. The replacement bridge was built by Alfred McAlpine Construction Ltd. The new Jackfield Free Bridge is a cable-stayed bridge built from steel and is notable for its tower, the top of which is reminiscent of the Eye of Sauron as depicted in the *Lord of the Rings* film trilogy. It was officially opened on 18 October 1994, although traffic had been using it since August. The first vehicle to use it was a 14-tonne steam roller as a tribute to the 'load test' that took place on the original Free Bridge. It carries the B4373 across it.

The Jackfield Bridge near Ironbridge in Shropshire. Its unusual-looking tower has drawn much attention. (Cheryl Billington)

Llanthony Road Bridge, Gloucester, Gloucestershire

Llanthony Road Bridge, or Llanthony Lift Bridge, is a steel lift bridge in Gloucester Docks which crosses the Gloucester and Sharpness Canal. It is the third such bridge on this site with a wooden swing bridge having been the first. It opened in 1794 when the docks were first built and carried the Llanthony Road across it. This was replaced by an iron swing bridge by the Midland Railway in the 1860s carrying rails to link to railway sidings on both sides of the docks. The current bridge was opened in 1972 and is operated by a bridge keeper. The bridge can only open between 9.00 a.m. and 6.00 p.m. and must open for any vessel passing through, from a narrowboat to a tall ship. It is usually lifted at least once a day. Only buses, taxis, cyclists and pedestrians are allowed to use the bridge.

The Llanthony Road Bridge in Gloucester Docks is a steel lift bridge which only lifts between the hours of 9.00 a.m. and 6.00 p.m. (Stephen Bunch)

Lower Kings Road Bridge, Berkhamsted, Hertfordshire

This colourful road bridge goes across the Grand Union Canal in Berkhamsted, Hertfordshire, next to lock gate 53. The Grand Union Canal opened in stages between 1793 and 1805 and joined London with Birmingham and Leicester, passing through Berkhamsted. Lower Kings Road, including the bridge across the canal, was opened in 1885 with the purpose of joining the railway station north of the canal with the town centre to the south. The railway station had been opened in 1875, replacing a previous station from 1837 further to the east. The need for a new road joining the second railway station with the town centre was met by public subscription, with the road costing just over £3,000. Also known as 'Canal Bridge' or 'Kings Road Bridge' with the number 140c, there is a former mill next to the bridge, which is now flats and offices. In 2020–21 the bridge underwent repairs and was colourfully painted in the style reminiscent of a typical canal narrowboat by Phil Speight, a traditional canal art painter. There are three oval-shaped aluminium panels on each side of the bridge showing various scenes, such as nearby Berkhamsted Castle where William the Conqueror was declared king in 1066. The centre one on the north side says 'Welcome to the Port of Berkhamsted' as further along the canal is the inland port of Berkhamsted. The paintings were commissioned by a partnership comprising Berkhamsted Citizens Association, Hertfordshire County Council and Osborne, a local company.

The panels on this colourful bridge over the Grand Union Canal at Berkhamsted in Hertfordshire were painted by the traditional canal artist Phil Speight.

A view over Magdalen Bridge looking towards Oxford city centre with the tower of Magdalen College at the far end.

Magdalen Bridge, Oxford, Oxfordshire

Magdalen Bridge in Oxford crosses two channels of the River Cherwell to the east of Oxford city centre. The current bridge was constructed between 1772 and 1790 by John Randall and was designed by John Gwynn of Shrewsbury. It has eleven arches of different dimensions in total. For each branch of the river there are three large semicircular arches with two smaller ones on each side. Right in the middle is a central elliptical arch which straddles the island area between the two river channels. If the river floods, water goes under this central bridge. Although traffic began using the bridge in 1778, work on a balustrade and other parts meant that the bridge wasn't actually completed until 1790. In 1882 the bridge was widened so that a 4-foot-gauge track for a tramway could be added. Today the bridge carries the A420 into the city centre from the east. Its name comes from Magdalen College on the west side of the bridge.

Matlock Bridge, Matlock, Derbyshire

The Matlock or Derwent Bridge in Derbyshire crosses the River Derwent in the town of Matlock. It consists of four pointed arches built of stone and is Grade II listed. It was

Matlock Bridge crosses over some rapids on the River Derwent in Derbyshire.

built in the 1400s as an alternative crossing of the river here to a ford. It was only 13 feet (4 metres) wide at first but was widened by around 18 feet (5.5 metres) in 1904 on the north-western side. It joins the town centre on its northern side with the railway station on its southern side. The area around the bridge is also known as Matlock Bridge. The road across the bridge is called Snitterton Road.

Mill Bridge, Bourton-on-the-Water, Gloucestershire

Bourton-on-the-Water is a Cotswolds village north of Cirencester, Gloucestershire. Since Victorian times it has been a popular tourist destination with attractions including a model village, a motoring museum and the River Windrush which flows through the centre of the village. Originally the river flowed to the south of the village, but in the sixteenth century it was redirected through the village to provide water power for three mills there. As a result, five low-lying, tiny bridges were built across the river. The first one to be built was Mill Bridge, built in 1654 from local Cotswold stone, replacing a ford that was there. It was originally known as 'Broad' or 'Big' bridge as motor vehicles can cross it despite its size. Four more bridges were built over the

The smallest bridge in this book, Mill Bridge over the River Windrush in Bourton-on-the-Water. It was built in 1654 after the river was diverted through the village.

years: High Bridge in 1756, Paynes Bridge in 1776, New Bridge in 1911 and Coronation Footbridge in 1953. Due to the picturesque river with the tiny bridges crossing it, Bourton-on-the-Water has been called the 'Venice of the Cotswolds'.

Mill Lane Bridge, Leicester, Leicestershire

Situated on the Mile Straight of the River Soar Navigation in Leicester is Mill Lane Bridge, south-west of the city centre. It connects De Monfort University on the east side of Mill Lane with the Bede Island area to the west. It is alternatively called Bridge No. 110 on this section of the Grand Union Canal. The canalized section was built through the city of Leicester in the nineteenth century in order to reduce the floods that frequently occurred there. The bridge was built in 1890 and is now Grade II listed. Its design is unusual in that it incorporates two different bridge styles. It is a braced arch bridge with two segmented iron arches on either side of the roadway, which is a box-girder type of bridge. There are also four stone octagonal turrets at the four corners of the bridge which have caps on top. There is another more or less identical bridge to the south of this one on Upperton Road with the only difference being the turrets, which are more pointed.

Mill Lane Bridge in Leicester with the buildings of De Monfort University in the background. (Mike Pollock)

Mythe Bridge, Tewkesbury, Gloucestershire

The Mythe Bridge to the west of Tewkesbury in Gloucestershire carries the A438 over the River Severn. It was opened on 12 May 1826 as a toll bridge. Its original design by George Moneypenny was of a three-arch bridge which was changed to a single-arch design by Thomas Telford when it was discovered that the ground in the river wouldn't take the weight of two pillars midstream. The bridge has a single arch with a span of 170 feet

The Mythe Bridge, west of Tewkesbury in Gloucestershire, was designed by Thomas Telford and crosses over the River Severn.

(52 metres) and is constructed from cast iron. The bridge cost around £35,000 (£3.5 million in 2019) compared with the original estimate of £20,700.

It was hoped that the bridge would attract long-distance traffic between Hereford and London, but the heavy tolls put off people from using it. In 1891 it was sold to the County of Gloucester and the tolls were abolished. The bridge became Grade II listed in 1952. Then in 1990 a 7.5-tonne weight limit was introduced along with single-file traffic. Two years later in 1992 the bridge was strengthened with steel plates which increased the weight limit to 17 tonnes, thus allowing fire engines to cross it.

Swarkestone Bridge, near Melbourne, Derbyshire

Swarkestone Bridge carries the A514 over the River Trent to the south of Derby in Derbyshire. It is considered to be the longest stone bridge in England due to its connected causeway which is continuous for a mile to the south of the village of Swarkestone. It is actually 1,304 yards (1.2 km) long and both the bridge and causeway are Grade I listed buildings, with the causeway also being a scheduled ancient monument. It is also known as the Long Bridge, The Stanton Causeway and The Swarkestone Causeway. The bridge is first mentioned in 1204 when it was called the Ponte de Cordy and is thought to have originally been constructed as a wooden bridge across the River Trent. Both the bridge and causeway were then rebuilt with stone. We know that the original bridge was destroyed in 1795 when flooding caused debris to crash into the bridge. The rebuilt bridge was

Part of the Swarkestone Bridge and Causeway over the River Trent in Derbyshire, which is considered to be the longest stone bridge in England. (Nic Clacy)

designed by Samuel Lester of Leeds and cost £3,550. This bridge is still standing today but is noticeably different to the causeway having been constructed of yellow sandstone whilst the causeway has lighter coloured stone.

The bridge is a strategic crossing point of the River Trent and as such has seen some skirmishes in history. During the English Civil War a battle took place here on 6 January 1643 when Royalists led by Sir John Harpur defended the bridge against Parliamentarians under Sir John Gill. Then in December 1745 in the Jacobite Rebellion of Bonnie Prince Charlie seventy of his men were sent to defend the causeway from English troops who had come up north to destroy the bridge. Prince Charlie's men reached the causeway first and after a standoff, the English troops withdrew. However, Bonnie Prince Charlie and his men were forced to go back to Scotland due to lack of numbers and eventually were defeated by the English army at the Battle of Culloden in 1746.

In the 1800s the causeway was widened from 9 feet (2.75 metres) to 19 feet (5.8 metres) and the parapets were replaced. In 1899 some of the brickwork in the medieval arches was removed and replaced with blue bricks. There is a 40-mph speed limit on the bridge and causeway as well as a 7.5-tonne weight limit.

Telford's Bridge, London Colney, Hertfordshire

Telford's Bridge carries the B5378 across the River Colne at London Colney near St Albans in Hertfordshire, the source of the river being just a few miles away at North Mymms Park. It originally carried the A6 trunk road until a bypass to the village was opened in 1959.

This colourful red-orange brick bridge at London Colney is called Telford's Bridge though it is not certain that he had anything to do with the design and building of it.

The bridge dates from 1774/75 and although it is thought to be connected to the engineer Thomas Telford, this is unlikely as he wasn't born until 1757. At the time of the bridge's construction, he would only have been seventeen or eighteen and still living in Scotland. However, other sources say he was responsible for the engineering of the bridge approaches and this is the origin of the bridge's name. The bridge is quite picturesque, being built of orange-red bricks and having seven tunnel arches, which increase in size towards the centre arch. When the river level is low, you can walk under them. The bridge parapet and railings were added in the twentieth century. The bridge underwent restoration work in 1998 and is Grade II listed.

Trent Bridge, Nottingham, Nottinghamshire

Trent Bridge is the main bridge over the River Trent in Nottingham and as such carries the A60 London Road from the south into the city centre. Several bridges have stood on this site dating from 920, including one known as Hethbeth Bridge. The present Trent Bridge dates from 1871. It was designed by Marriott Ogle Tarbotton and built by Benton and Woodiwiss of Derby, with the Derbyshire iron makers Andrew Handyside providing the cast/wrought-iron girders. It cost £30,000 to build. The bridge has three main arch spans of 100 feet (30 metres) in length, with a width of 40 feet (12 metres) between the parapets. From 1924 to 1926 the bridge was widened to 80 feet (24 metres) by the Cleveland Bridge & Engineering Co. It is Grade II listed and gives its name to the nearby Nottinghamshire County Cricket ground. Also, near to the bridge are the football grounds of Nottingham's two football league clubs, Nottingham Forest and Notts County.

Trent Bridge in Nottingham gives its name to the nearby cricket ground of Nottinghamshire County Cricket Club.

Wallingford Bridge, Wallingford, Oxfordshire

Wallingford Bridge crosses the River Thames in Oxfordshire between Wallingford and Crowmarsh Gifford. It is 900 feet (270 metres) long and consists of nineteen arches. It dates from at least 1141 when it is first mentioned in written documents. Back then it was a 12-foot-wide (3.6 metre) bridge built of stone, with nineteen spans. Only four of these are still standing, having been incorporated into the current bridge. In 1530 after the closure of nearby Holy Trinity Priory during the Dissolution of the Monasteries the bridge was substantially repaired using stone from the priory. Five of the arches were rebuilt.

When Wallingford Castle was besieged by Parliamentary forces in the English Civil War four of the bridge's arches were removed and timber drawbridges were put in their place. It wasn't until 1751 that the arches were rebuilt with stone once more. In 1810–12, following a significant flood in 1809, the bridge was partially rebuilt. Three wider arches were added as well as a toll house, parapet and balustrade to a design by John Treacher. The bridge was also widened on the north side by 7 feet (2 metres). It was the main crossing of the River Thames here until 1993 when Winterbrook Bridge was built to the south as part of the southern Wallingford bypass. Due to the narrowness of the bridge a traffic-light system is in place to allow single-file traffic pass. It has been a scheduled monument since 1949.

The ancient stone bridge at Wallingford in Oxfordshire crosses over the River Thames and dates from the 1400s.

Welsh Bridge in Shrewsbury gets its name as it is nearer to the Welsh border than the other main bridge over the River Severn here, English Bridge.

Welsh Bridge, Shrewsbury, Shropshire

Welsh Bridge in Shrewsbury, Shropshire, takes the A458 at Frankwell over the River Severn on the north-western side of the town. Although the Welsh border is several miles away, it was given this name to distinguish it from the other bridge across the Severn in Shrewsbury, the English Bridge.

Originally there was a medieval bridge here known as St George's Bridge. It was replaced by the present stone bridge which was built between 1793 and 1795 by John Carline and John Tilley, who also designed it. The bridge is a masonry arch viaduct and is a Grade II listed building. It is 266 feet (81 metres) long and 39 feet (12 metres) wide. It has five arches constructed of Grinshill sandstone. The central arch is 46 feet 2 inches (14 metres) across, whilst the remaining four arches are 43 feet 4 inches (13 metres) across. The bridge cost £8,000 to build.

Whitney-on-Wye Toll Bridge, Herefordshire

Whitney-on-Wye Toll Bridge carries the B4350 south from its junction with the A438 and crosses the River Wye in Herefordshire. It is unusual in that it is essentially two different bridges joined together, with wood in the centre section and stone at either end. The first bridge was built here in around 1780 and had five stone arches. This bridge and two more were destroyed by floods, the last one being in 1795. The damaged part of the bridge was rebuilt in 1797 using a different design with a wooden beam and trestle design on three of

Whitney-on-Wye Toll Bridge is the only operating toll bridge to be featured in this book. It is unusual as it is built partly out of wood and partly out of stone. (Stephen Bunch)

the spans. It also had wooden decking built into it. In 1992–93 around £300,000 was spent on reconstructing the bridge. The bridge and toll house are both Grade II listed. Tolls are still charged with an estimated tax free income of £2,000 a week in 2012.

Worcester Bridge, Worcester, Worcestershire

Worcester Bridge crosses the River Severn in the centre of Worcester, Worcestershire. It is a five-arch stone bridge constructed from red sandstone and was designed by John Gwynn, who had been involved with the designs of Magdalen Bridge in Oxford and English Bridge in Shrewsbury. Previously there had been a fortified bridge here dating from the

Worcester Bridge crosses over the River Severn with Worcester Cathedral in the background.

fourteenth century. The present Worcester Bridge was opened in 1781, though this was widened in 1931–32. This newer bridge was officially opened by Edward, Prince of Wales in October 1932. It now carries two lanes of traffic in each direction and links four major roads: the A38, A44, A443 and A449. The bridge is susceptible to flooding quite often, and when this happens traffic is diverted across the Carrington Bridge to the south and the Holt Bridge to the north.

Workman Bridge, Evesham, Worcestershire

The Workman Bridge in Evesham is a three-arch segmental stone bridge over the River Avon in Worcestershire. It is named after Henry Workman, a local businessman and former mayor of Evesham. It was opened on 12 March 1866 as a replacement to the medieval bridge that had been on the site for several hundred years. It connects Evesham on the north side of the river with Bengeworth on the south side. It was restored in 1976 and is now Grade II listed. There are two commemorative plaques on the bridge. The first commemorates Henry Workman and his involvement in getting the bridge built, with the date of the opening as 1866. The second gives the date of the erection of the bridge as 1856, with James Samuel being the engineer and James Taylor being the contractor.

Workman Bridge in Evesham, which crosses the River Avon in Worcestershire, is named after a former mayor of the town, Henry Workman.

Wye Bridge, Hereford, Herefordshire

The Wye Bridge, or Old Wye Bridge as it is also known, spans the River Wye in Hereford. There was a wooden bridge here dating from Saxon times, but this was replaced with another timber bridge around 1100. The present sandstone bridge was built in 1490 to replace the previous bridge which had become dilapidated. It had a gatehouse built on it at the south end. During the English Civil War, the bridge was damaged during the Siege of Hereford in 1645. In 1826 the bridge was widened on the eastern side and evidence of this can be seen on the third arch on the Hereford side of the bridge. It has six arches and is Grade I listed. It carries St Martin's Street across the river. Most traffic now uses the nearby Greyfriars Bridge.

The River Wye in Hereford gives its name to the fifteenth-century bridge which crosses over the river here.

4

Northern England

Ashopton Bridge/Viaduct, Ladybower Reservoir, Derbyshire

The Ashopton Bridge carries the A57 Manchester to Sheffield Road across the Ladybower Reservoir in Northern Derbyshire. In the 1930s a decision was made to flood the valley of the River Derwent in Derbyshire to create a vast reservoir which would provide water for the nearby conurbations of Sheffield and Manchester. This meant that some of the villages in the valley would need to be evacuated and then demolished. These were Ashopton and Derwent. Work to create the reservoir started in 1935 but was delayed when the Second

This unusual bridge taking the A57 over the Ladybower Reservoir in Derbyshire looks like a bridge but was in fact built as a viaduct and is now mostly covered by water. (Photo © Peter McDermott, CC by-SA 2.0)

World War began in 1939. Two viaducts – the Ashopton and Ladybower – were built to carry the A57 and A6013 roads over the reservoir. They were built by Holloways of London who used a steel frame set in concrete to create the viaducts. As the water levels rose the viaducts soon started to look like bridges and this is how they are seen and described today. The reservoir was actually completed in 1943 and opened by King George VI on 25 September 1945. Occasionally in times of drought the water level drops and signs of the villages can be seen, as in 1976 and 2018.

County Bridge, Sowerby Bridge, Calderdale, West Yorkshire

The bridge over the River Calder in West Yorkshire is known as County Bridge and carries the A58 across it. A bridge has been here since at least 1314 and was mentioned in documents when the townspeople did not keep it in good repair. It connected the settlement of Sowerby on the west side with other villages on the east side and lead to the town getting its name. By 1517 the original wooden bridge had been rebuilt in stone, having three round arches as well as piers and triangular cutwaters. The bridge was widened in 1632 and in 1875. It came to be known as County Bridge in 1673 when the upkeep of the bridge was handed over to local Justices of the Peace following flood damage. An iron superstructure has been added to the stonework on the western side in more recent years. It has a latticed base with a balustrade of round-ended panels. It is a scheduled monument.

County Bridge at Sowerby Bridge in West Yorkshire crosses the River Calder with parts of it dating from 1517.

Dunsop Bridge is the road bridge deemed to be nearest to the geographical centre of Great Britain. (Alan Morris)

Dunsop Bridge, Dunsop, Lancashire

Dunsop Bridge in Lancashire gives its name to the village in which it is situated. This is at the confluence of the River Dunsop and the River Hodder, which are both tributaries of the River Ribble which then flows through Clitheroe and Preston on its way to the Irish Sea. The bridge has one single sandstone arch and was built in the early nineteenth century. It is a Grade II listed structure. It is included in this book as it is believed to be the nearest bridge to the geographic centre of Great Britain. According to the Ordnance Survey this is said to be at Whitendale Hanging Stones, near Brennand Farm, approximately 4 miles (7 km) north of the village.

East Marton Bridge, East Marton, North Yorkshire

The bridge over the Leeds and Liverpool Canal at East Marton village in North Yorkshire is unusual in that it consists of two bridges, with one on top of the other. Known as 'Bridge # 161', the upper bridge carries the A59 through the village, between Skipton and Gisburn.

This unusual 'double bridge' crosses over the Leeds and Liverpool Canal at East Marton in North Yorkshire. (Ian Watts)

Originally a small packhorse bridge was built across the canal here in around 1790 when the canal was being built. Then when the A59 road was opened here much later, it was deemed easier to build the second bridge on top of the original bridge to keep the road level. It is known as the double-arched bridge.

Eden Bridge, Carlisle, Cumbria

The Eden Bridge (or New Bridge) to the north of Carlisle city centre in Cumbria carries the A7 over the River Eden out of Carlisle northwards. Timber bridges were built across the river here as far back as the twelfth century and then a stone bridge in the sixteenth century. The present Eden Bridge replaced this and was opened in 1815, having been designed by Sir Robert Smirke. It is an arch bridge with five arches, each being 65 feet (20 metres) long, and is made of sandstone. It has four lanes for road traffic, being widened from 33 feet (10 metres) to its present width of 70 feet (21 metres) in 1933. For many years it was the nearest river crossing to the Solway Firth until 2012 when the A689 western bypass for Carlisle was opened with a new bridge nearer to the sea.

Eden Bridge carries the A7 road north out of Carlisle and into Scotland. (Photo © Bobby Clegg, CC by-SA 2.0)

Elvet Bridge, Durham, County Durham

The Elvet Bridge in Durham is the oldest bridge to cross the River Wear, dating from the late twelfth century. It connects the east side of Durham with the cathedral peninsular. There were originally chapels at either end of the bridge. Today there is just one of the

The Elvet Bridge connects the cathedral area with the east part of Durham and dates from the twelfth century.

seven original arches with the rest dating from the thirteenth century, around 1225 when most of the bridge was reconstructed. Back then there were fourteen pointed arches, but today there are ten that can be seen, whilst the other four have been covered by buildings. In 1805 the bridge was widened from 17 feet (5 metres) to 31 feet (9.5 metres). The bridge is Grade I listed and is pedestrianised for most of the day, with service vehicles allowed to use it during certain hours.

Ennerdale Link Bridges, Kingston upon Hull, East Riding of Yorkshire

The Ennerdale Link Bridges are two bridges which carry the A1033 road, Raich Carter Way, a dual carriage way, over the River Hull on the northern edge of Kingston upon Hull in the East Riding of Yorkshire. They were opened on 3 April 1997 as part of the Ennerdale Link development at a total cost of £30 million. They are bascule link bridges that move in a similar way to a drawbridge, so that they can be quickly lifted up to allow river traffic to pass underneath. Originally, it was planned to build a tunnel under the river here, and work started in 1991. However, after a nearby dam burst, the ground here

The unusual-looking Ennerdale Link Bridges north of Kingston upon Hull. (Photo © Paul Harrup, CC by-SA 2.0)

was deemed to be too unstable to merit a tunnel and work was stopped in 1993. They are similar in design to the Stoneferry Bridges which are situated 3 miles further downstream into Hull. Both sets of bridges were installed by Humberside County Council and are painted in a blue colour scheme. The Ennerdale Bridges weigh 800 tonnes in total and are rarely opened nowadays, apart from when allowing occasional barges and pleasure boats to pass underneath.

Grosvenor Bridge, Chester, Cheshire

The Grosvenor Bridge carries the main A483 over the River Dee, south of Chester towards Wrexham. It is named after the Duke of Westminster's family name, Grosvenor, and was designed by the architect Thomas Harrison. The bridge is a single-span arch stone bridge 200 feet (61 metres) long and 60 feet (18 metres) high and was built between 1827 and 1833. The bridge is constructed of different coloured sandstone blocks, mainly pale red and beige which gives the effect of making the bridge look pink in colour. It was formerly opened on 17 October 1832 by Princess Victoria of Saxe-Coburg-Saalfeld and her daughter Princess Alexandrina Victoria of Kent (later to become Queen Victoria). The bridge wasn't quite finished then and wasn't completed until November 1833. It was the longest single-span stone arch bridge in the world when it was first opened and this honour lasted for thirty years. It is now a Grade I listed building.

It was built partly to replace the nearby Old Dee Bridge which was quite narrow and partly to cash in on road traffic going between London and Holyhead. A toll was charged to pay for the construction costs of £50,000. This was counterproductive as traders were put off from using the bridge. The toll was finally ended in 1885 when the Chester Corporation took over the maintenance of the bridge.

Grosvenor Bridge in Chester takes the A483 over the River Dee and dates from 1833.

Hulme Arch Bridge is seen as a gateway into Manchester from the south. (Photo © Peter McDermott, CC by-SA 2.0)

Hulme Arch Bridge, Hulme, Greater Manchester

The Hulme Arch Bridge, or Stretford Road Bridge, in Hulme, Manchester, takes the A5067 Stretford Road over the A5103 Princess Road, a major arterial road south from Manchester city centre. It was opened in 1997 as part of the regeneration of the Hulme area and reconnected Stretford Road which had been cut into two parts when the Princess Road was opened in 1969. The bridge was formally opened on 10 May 1997 by the then manager of Manchester United, Alex Ferguson. A closed competition had been held for a new bridge which would reunite both truncated parts of Stretford Road and be seen as a gateway into Manchester. It was won by Chris Wilkinson Architects, who referenced their design on the Gateway Arch in St Louis in the USA. The structural engineer was Ove Arup & Partners. The bridge deck is supported by twenty-two steel cables connected to an 82-foot-high (25 metre) steel arch which crosses over from one side of the bridge deck to the other. The bridge deck is 160 feet (50 metres) long and is made up of three 56 × 56-foot (17 × 17 metre) concrete and steel segments. The bridge has won several awards for its design.

Irwell Street Bridge, Salford/Manchester, Greater Manchester

The Irwell Street Bridge across the River Irwell connects Irwell Street in Salford on the north side with New Quay Street in Manchester on the south side. It opened in 1877 and was designed to help ease the pressure of coal traffic going to the Lancashire and

The Irwell Bridge crosses the River Irwell between Manchester and Salford and was built to help transport coal between the two cities.

Yorkshire goods stations using other nearby bridges. Today it carries the A34 across the Irwell and as both roads were not quite square the bridge had to be built slightly askew. It is a bowstring arch truss bridge with iron girders and lattice-work cross bracing. It was designed by James Gascoigne Lynde who was the City Surveyor to Manchester Corporation. The two main girders are 130 feet (40 metres) long, whilst the remaining ones are 127 feet (39 metres) in length. The width of the roadway between the parapets is 48 feet (15 metres). It is Grade II listed.

Lady's Bridge, Sheffield, South Yorkshire

Lady's Bridge is the oldest bridge across the River Don in the City of Sheffield. It is located to the north of the city centre, linking the Wicker in the north with Waingate in the south. The bridge dates from 1485 when the Vicar of Sheffield, Sir John Plesaunce, and William Hill, a master mason, agreed to build a stone bridge near to Sheffield Castle, which is now demolished. The bridge had five arches, and a small chapel dedicated to the Virgin Mary was built onto it. Eventually the bridge came to be known as 'Our Lady's Bridge' due to this connection. It was 14 feet (4.3 metres) wide and could only be used by pedestrians due to there being steps at either end. With the Dissolution of the Monasteries under

Lady's Bridge in Sheffield is the oldest bridge to cross the River Don in the city and has survived nearly being destroyed by floods in both 1864 and 2007.

Henry VIII, the chapel became a warehouse used for storing wool and then eventually an almshouse. It was demolished in 1760 when the bridge was widened. In both 1864 and 2007 during heavy floods that hit Sheffield the bridge was close to collapse but survived on each occasion.

Lendal Bridge, York, North Yorkshire

The Lendal Bridge crosses the River Ouse in York, North Yorkshire, connecting the city centre with the railway station and the western part of the city. It replaced an earlier rope ferry on this site. The current bridge dates from 1863 and was designed by Thomas Page, who also designed Westminster Bridge in London. There was a previous bridge here which collapsed during construction. Work had started in 1860 under the guidance of William Dredge, but its collapse resulted in the deaths of five workmen. A new bridge, the current Lendal Bridge, replaced this one and was opened in 1863. It was constructed of cast iron and has a single span of 175 feet (53 metres). It connects two medieval towers on either

Lendal Bridge crosses over the River Ouse in York and gets its name from one of the two towers built onto it.

side of the river. The Lendal Tower, from which the bridge gets its name, is on the east bank, and the Barker Tower is on the west bank. The bridge has been used as the location for two television series: *Damon and Debbie* (1987), a spin-off of the soap opera *Brookside*, and *Gunpowder* (2017), a BBC historical drama.

Old Dee Bridge, Chester, Cheshire

The Old Dee Bridge over the River Dee in Chester connects Chester city centre with the village of Handbridge. The present bridge is believed to date from the Middle Ages, but it is thought that there were other bridges here during Roman times when Chester was the Roman fortress Deva. The bridge is constructed of red sandstone, which was mined locally, and has seven arches, each one being different in size. It is believed the present bridge dates from 1387. Originally there was a drawbridge on the south side of the bridge, but this was replaced by an arch. There was also a tower on the bridge, which was built between 1399 and 1407, along with a gatehouse, which was demolished in 1781. The bridge was widened in 1825–26 by Thomas Harrison and shortly after this in 1827 work was started on the Grosvenor Bridge a few yards downstream as a better alternative to this medieval bridge. Today the bridge is still used by road traffic, though it has a single carriageway, which is controlled by traffic lights.

Old Dee Bridge in Chester dates from 1387 and is built from local red sandstone.

Redheugh Bridge, Newcastle upon Tyne, Tyne and Wear

The Redheugh Bridge, which crosses the River Tyne to the south-east of Newcastle upon Tyne, is the third bridge to have been built at this site. It is a pre-stressed three-span concrete bridge and is 2,943 feet (897 metres) long. It has a central span which is 520 feet (160 metres) long, with two side spans which are 330 feet (100 metres) in length. It is 52 feet (15.6 metres) wide and carries two lanes of traffic in each direction. It was opened on 18 May 1983 by Diana, Princess of Wales and cost just over £15 million, being built by Edmund Nuttall Ltd. It was built to last 120 years.

The Redheugh Bridge over the River Tyne in Newcastle upon Tyne was opened in 1983 and designed to last 120 years.

Selby Toll Bridge over the River Ouse stopped charging a toll in 1991 but still retains 'toll' in its name.

Selby Toll Bridge, Selby, East Riding of Yorkshire

The Selby Toll Bridge over the River Ouse in the East Riding of Yorkshire is a swing bridge which was opened in 1970. It was designed as a replica bridge to the previous wooden swing bridge which had been on this site since the 1790s. It was the nearest crossing point to the sea of the River Ouse for many years until the Boothferry Bridge was opened in 1929. It is still called the Selby Toll Bridge despite the fact that tolls were formally abolished on 21 September 1991 after many years of constant gridlock in the town caused by queuing at the bridge. It used to carry both the A19 and A63 roads across the river until the A63 bypass road was opened in 2004 to the south of the town. It is one of three swing bridges over the River Ouse in Selby.

Skeldergate Bridge, York, North Yorkshire

The Skeldergate Bridge over the River Ouse in York was opened on 1 January 1888. At first only pedestrians could use it and it was only opened to other traffic a few months later.

The bridge is made from iron and has many intricate details including six pointed stars, trefoils and a large white rose of York. It was designed by Thomas Page, who had also designed the nearby Lendal Bridge as well as Westminster Bridge in London. When he died early on in the development, his son, George, took over. It was designed so that the north-easternmost span of the bridge could open for tall ships on their way to the quaysides upstream. The bridge last opened in 1975. It was a toll bridge until 1914.

Skeldergate Bridge in York which is often subject to flooding from the River Ouse.

Skew Bridge, Rainhill, St Helens, Merseyside

This road bridge over the railway line between Liverpool and Manchester at Rainhill in Merseyside dates from June 1829 and is the oldest skew bridge over a railway line in the world. A skew bridge is a bridge which is built at a different angle than the usual 90-degree perpendicular angle that is seen in the majority of road bridges crossing over a river

The Skew Bridge at Rainhill over the Liverpool to Manchester railway line is the oldest skew bridge over a railway line in the world, dating from 1829.

or railway. This bridge crosses over the railway line at an angle of 34 degrees as it was impossible to realign either the road or the railway at this point. It is sometimes known as the George Stephenson Skew Bridge after the engineer who designed it, as he was the Principal Engineer for the whole line. The bridge cost £3,735 and is now Grade II listed.

Rainhill is famous as the place where the Rainhill Trials took place in 1829 to find the best railway engine to be used in service on the first passenger railway line in the world between Liverpool and Manchester. It was won by George's son, Robert Stephenson, with his *Rocket* locomotive. The bridge is constructed of sandstone and the engineer who built it was George Findlay. The bridge was widened in 1963 to cope with increased road traffic. The original milestone indicating distances to Liverpool, Prescot and Warrington on the bridge was placed on the wrong side of the bridge and it wasn't until 2005 that this error was corrected and the milestone put on the correct side.

Sutton Weaver Swing Bridge, Frodsham, Cheshire

The Sutton Weaver Swing Bridge near Frodsham in Cheshire carries the A56 across the River Weaver Navigation, a tributary of the River Mersey. It was designed as a swing bridge to allow boats to travel upstream to Northwich. It was opened in 1926 and replaced

The distinctive Sutton Weaver Swing Bridge over the River Weaver Navigation at Frodsham dates from 1926.

an earlier swing bridge which had been operating here since 1872. It was designed by Colonel John Arthur Saner and built by the company of Joseph Parks and Son of Northwich. It is constructed of steel girders in a lattice design and is painted dark grey with white edging over the top. The swing bridge is mostly supported by a float, with the rest being supported by 114 rollers. The bridge is able to swing through an arc of 100 degrees in around a minute via a wire rope turned by an electric motor.

Tees Barrage, Stockton-on-Tees, County Durham

The Tees Barrage is a bridge structure over the River Tees between Stockton-on-Tees and Thornaby in Teesside. It is made up of a road bridge, a footbridge, a fish pass, a barge lock and the river barrage itself. It also has a specially built white-water course next to it. It opened on 22 April 1995 and was designed by Ove Arup and the Napper Partnership. Work on the barrage started in November 1991 with Tarmac Construction winning the contract to build it. The River Tees was diverted around the construction site which was 'in the dry' to make construction work easier. On completion the river was put back on its natural course. The waters above the barrage are held at average high-tide level in order to provide facilities for various water sports including jet skiing, rowing, canoeing and dragon boat racing.

The Tees Barrage at Stockton-on-Tees is unusual in that it is not just a road bridge, but also a footbridge, a fish pass, a barge lock and a river barrage all rolled into one.

Thelwall Viaduct, Warrington, Cheshire

The Thelwall Viaduct on the M6 motorway to the east of Warrington in Cheshire consists of two separate viaducts carrying the roadways of the motorway in each direction. It is situated between junctions 20 and 21 of the M6 and crosses both the River Mersey and the Manchester Ship Canal.

The first bridge was opened in July 1963 when the early stages of the M6 motorway were being constructed between Preston and Warrington. It had three lanes in each direction but no hard shoulders and provided relief for traffic using the A49 and A50 roads in and around Warrington. It was actually the longest road bridge in Great Britain for several months until the Forth Road Bridge was opened.

Due to its popularity and steep approaches, the traffic jams on the local roads were now also occurring on the Thelwall Viaduct. With the widening of the M6 in the 1990s, a new bridge was built alongside the 1963 bridge at a total cost of £52 million. This was opened in 1995 and meant that the crossing had four lanes in each direction as well as hard shoulders. The older bridge is 4,414 feet (1,345 metres) long and carries the northbound carriageway; the southbound newer bridge is 4,593 feet (1,400 metres) long. The span over the waterways below is 336 feet (102 metres). Between 2002 and 2005 remedial work took place to replace 148 roller bearings which had failed. This resulted in long tailbacks until the work was completed.

Thelwall Viaduct is actually two separate bridges carrying the M6 motorway across the River Mersey and the Manchester Ship Canal. (Richard Woolham)

Victoria Bridge, Leeds, West Yorkshire

Victoria Bridge crosses over the River Aire in central Leeds to the south of Leeds railway station. It was opened in January after nearly twenty years of planning arguments and floods delaying its construction. Previously the river had been forded here by local residents, but a footbridge was first built here in 1819. However, due to objections from the Leeds and Liverpool Canal Corporation it was soon taken down. This was followed by a 'Union Bridge', a temporary wooden structure which opened in July 1829. Finally in 1836 Royal Assent was given to build a permanent bridge here. It was decided to call it the Victoria Bridge after the Princess Victoria who would ascend the throne in 1837. The contractors were George Leather & Son of Leeds and the structure would be a single-span bridge 80 feet (24 metres) across. Work began in February 1837 with the foundation stone being laid on 10 May 1837. However heavy rain and floods in December 1837 caused damage to the bridge which led to further delays and the need for more funds to be raised. It was eventually completed in December 1839 and finally opened to pedestrians and horse-drawn traffic. The bridge was tolled until 1867.

Victoria Bridge in Leeds crosses the River Aire just to the south of Leeds railway station.

Wainwright Bridge, Blackburn, Lancashire

Wainwright Bridge in Blackburn, Lancashire, is part of the Blackburn Orbital Road which gives Bolton Road a link to Barbara Castle Way, crossing over Freckleton Street. The bridge was opened in June 2008 by the local MP Jack Straw, and cost £12 million. It has a single-span bowstring arch made from a steel/concrete composite. It is 260 feet (79 metres) in length. It has tubular twin arches on either side of the deck and is wide enough to carry a dual carriageway across it. The bridge was named after local Blackburn man Alfred Wainwright, famous for his books on fell walking around the Lake District. His name was chosen after a vote in the local newspaper. It was engineered by Balfour Beatty and designed by Yee Associates. In 2021 the bridge was painted in a mixture of blue and white, the colours of the local football team, Blackburn Rovers FC, after it had previously been painted a pale blue colour.

The Wainright Bridge in Blackburn is named after the famous Lake District fell walker Alfred Wainwright and opened in 2008. (Ian Park)

Warrington Bridge, Warrington, Cheshire

Warrington Bridge crosses the River Mersey at the south end of Warrington town centre. It is a single-arch bridge built out of concrete and has eight parabolic arch ribs. It is 134 feet (40 metres) long and 80 feet (24 metres) wide. It was built between 1909 and 1915 by Alfred Thorne & Sons and is the sixth bridge to be built here. For many years this was the lowest bridge crossing of the River Mersey until the Runcorn Swing Bridge was opened in 1905. In 1986 a new bridge across the Mersey was opened nearby in a bid to ease congestion on Warrington Bridge. The following year it underwent substantial refurbishment. The bridge links to Warrington town centre via Bridge Street.

Warrington Bridge is made from concrete and crosses over the River Mersey.

5

Scotland

Albert Bridge, Glasgow

The Albert Bridge crosses the River Clyde in Glasgow between Saltmarket by Glasgow Green in the north and Crown Street to the south. It was completed in 1871 and named after Prince Albert, Queen Victoria's late husband. It is the fifth bridge on this site, with the previous 1794 bridge, designed by Robert Stevenson, damaged by floods and subsequently being demolished in 1868. Its current replacement was designed by Bell & Miller and is constructed mainly of cast iron with concrete piers and abutments. There are three arches, all made of wrought iron. It is highly decorated with the arch ribs showing the coats of arms of Queen Victoria, Prince Albert and various Glaswegian corporations. There are also stone pillars which support the parapet which are decorated with Queen Victoria and

Albert Bridge in Glasgow is noticeable for its distinctive green colour and was opened in 1871.

Prince Albert medallions created by sculptor George Robert Ewing. The Albert Bridge is one of several bridges that cross the River Clyde in this vicinity including Victoria Bridge, the City Union Bridge, Glasgow Bridge, King George V Bridge and Rutherglen Bridge. It has a distinctive green colour and was extensively refurbished during 2013–15. The following year Prince Edward visited the bridge to see the completed works.

Clyde Arc, Glasgow

The Clyde Arc is a steel arch bridge over the River Clyde in the west end of Glasgow. It connects Finnieston Street on the north bank near to the Scottish Exhibition Centre (SEC) with Govan Road by Glasgow Science Centre and Pacific Quay on the south side. It is noteworthy due to the curve of the arch which crosses the bridge deck from one side to the other and because of this locals refer to the bridge as the 'Squinty Bridge'. The bridge was built to help regenerate the areas on both sides of the river here and to provide better access to Pacific Quay. It was officially opened on 18 September 2006 by the leader of Glasgow City Council, Steven Purcell. Construction work had only started in May 2005 with the works being completed in April 2006. It was designed by Halcrow Group and built by BAM Nuttall at a cost of around £20 million.

The bridge has a main span of 315 feet (96 metres), with a width of 72 feet (22 metres). The central navigation height at mean water height is 18 feet (5.4 metres). It carries four lanes of traffic across it with two lanes being reserved for buses, taxis and cycles. There are also pedestrian and cycle paths. It was designed to last 120 years and to cope with a possible light rapid transit system (light railway scheme) or even a tram system in future years. The bridge was shut for six months in 2008 when one of the bridge's fourteen hangers (or supporting cables) snapped. This was caused by defective fork connectors on the hanger and these had to be replaced. Before its official name was revealed, the bridge was known as the Finnieston Bridge.

The Clyde Arc on the west side of Glasgow was opened in 2006 and has been nicknamed 'Squinty Bridge' by locals. (Allan Heron)

General Wade's Bridge, Aberfeldy, Perth and Kinross

Wade's Bridge (or Tay Bridge) at Aberfeldy carries the B846 across the River Tay in Perth and Kinross. It dates from the eighteenth century and was one of a series of military roads and bridges built under the leadership of Lieutenant General George Wade, the commander in chief of George II's forces in Scotland. Between 1725 and 1737 over 250 miles of road and forty bridges were built by Wade's soldiers. This was mainly due to the fears of more Jacobite risings after ones in 1689 and 1719, and the 1715 Rebellion by Bonnie Prince Charlie. With plenty of good-quality roads and bridges in place, it would be easier to get soldiers and military equipment quickly to where they were needed.

The bridge has an unusual neoclassical design by architect William Adam with four obelisks over the middle arch, which has a raised parapet giving the impression of steps up to the central section. In reality, they hide a 16-foot-wide (4.9 metre) humpback roadway. There are also pyramid structures at each end of the five-arch bridge.

Work on the bridge started on 23 April 1733 when General Wade laid the foundation stone. The limestone used in the bridge came from the nearby Bolfracks estate, but inevitable delays meant the bridge wasn't opened until October 1734, with the official opening ceremony taking place the following summer. The total cost of the bridge was

General Wade's Bridge is the only surviving bridge still in use that was built under the command of General Wade. It was opened in 1784. (Pete Wing)

around £4,000. It measures 368 feet (112 metres) across, with a central span of 60 feet (18.3 metres) and four smaller aches of approximately 35-foot (10.7 metre) spans. There are two plaques on the bridge celebrating its engineering achievements. Due to its narrow carriageway and the central humpback section, traffic lights are used to control the flow of traffic with a 20-mph speed limit. The bridge is the only one of General Wade's bridges still in everyday use today and is a Grade A listed building. At the northern end of the bridge is the Black Watch Memorial, built in 1887 to mark Queen Victoria's Golden Jubilee. It celebrates the first muster of the Black Watch Regiment (42nd Royal Highlanders), which took place in a field on the north bank of the river, opposite the memorial.

New Bridge, Stirling, Stirling Council Area

New Bridge (or William IV Bridge) carries the A9, Causewayhead Road across the River Forth to the north of Stirling city centre. It was built between 1829 and 1833 under the engineer Robert Stevenson and was called 'New Bridge' as it superseded the previous stone bridge nearby which was humpbacked and very narrow. This is now known as 'Old Bridge' or 'Auld Bridge' and is no longer open to traffic. New Bridge was opened in May 1833 and has the alternative names of William IV Bridge or the River Forth Bridge. Until the Kincardine Bridge was opened in 1936, New Bridge was the nearest crossing point of the Forth to the sea. New Bridge is a five-span bridge of rustic ashlar stone construction, complete with segmental arches and rounded cutwaters. It is 33 feet (10 metres) wide and has a 60-foot-wide (18 metre) central arch. With the opening of the M9 motorway in the early 1970s to the west of Stirling, most of the traffic that had been using New Bridge as a through route started using this route instead.

New Bridge at Stirling was opened in 1833 to replace the nearby 'Old Bridge' dating from the 1400s. (Allan Heron)

North Bridge in Edinburgh is unusual in that it goes over the top of Waverley railway station sloping downwards from south to north.

North Bridge, Edinburgh

North Bridge is in the centre of Edinburgh and links Princess Street and Edinburgh's New Town in the north with Edinburgh Old Town in the south. It is different in that it crosses over both railway lines and a road, as opposed to water as with the majority of the bridges in this book. It carries the A7 above Edinburgh Waverley railway station and Market Street before it meets the High Street by the old Tron Kirk. It is unusual in that it inclines downwards at an angle from the south to the north. Originally work was started on a bridge here in 1763, being designed by William Mylne, but in 1769 before it was finished the bridge collapsed, killing five people. It took three more years before it was finally completed in 1772. However, this bridge was superseded by the present bridge which was built between 1894 and 1897 by Sir William Arrol & Co. and engineered by Blyth and Westland when Waverley station was extended.

The bridge has four lanes for most of its length (two being used for buses) and comprises three steel arches on stone piers with cast-iron decoration. It is 525 feet (160 metres) long and 75 feet (23 metres) wide. It has three spans of arched girders which are 175 feet (53 metres) long. A large war memorial is situated on the east side. Since 2021, the bridge has been undergoing a major refurbishment. It was expected that this would be completed within nine months, but due to the poor condition of the concrete on the bridge the time frame has been extended several times. At the time of writing, it is expected to be completed by spring 2026, with an estimated total cost of £85 million.

Renfrew Bridge, Renfrew, Renfrewshire

The Renfrew Bridge across the River Clyde connects Renfrew on the south side with Clydebank and Yoker on the north side. The bridge is a cable-stayed moveable twin-leaf

One of the newest bridges in this book, Renfrew Bridge to the west of Glasgow opened in May 2025. (Allan Heron)

swing bridge and is designed to open easily for ships using the river. It is built from steel and is just over 600 feet (184 metres) in length. It was built by the civil engineers Graham as part of the Clyde Waterfront and Renfrew Riverside Project to boost economic growth in this region. The two main sections of the bridge were constructed in the Netherlands and were then floated down the River Clyde and installed in late 2024. The bridge itself cost around £50 million, with the total cost of the whole project being £117 million. It was opened on 8 May 2025 with a ceremony in which schoolchildren from both sides of the bridge met in the middle. Traffic and pedestrians were able to use the bridge the following day.

Rumbling Bridge, Perth and Kinross

This is the name of both a village and a 'double bridge' in central Scotland, around 10 miles south-west of Perth. It carries the A823 over the River Devon in a steep gorge and it is unusual, in that it is made up of two bridges, one on top of the other. The lower and older bridge dates from 1713, with William Gray being named as the builder. It is 22 feet (6.7 metres) long and 11 feet (3.4 metres) wide, and 86 feet (26 metres) above the average water level. The upper bridge dates from 1816 and was built to make the road crossing

Rumbling Bridge in central Scotland is one bridge built on top of another. (Allan Heron)

the river gorge more level. It is 34 feet (10 metres) above the lower arch and 120 feet (37 metres) above the river. The bridge gets its name from a rumbling sound that can be heard at its lowest level.

Smeaton's Bridge, Perth, Perth and Kinross

The road bridge across the River Tay in Perth, known as Smeaton's Bridge, was built in 1771 and is named after the engineer John Smeaton who is credited with building it; it is also known as Perth Bridge or the Old Bridge. It carries the A85 road across it, linking Perth city centre on the west side with the district of Bridgend on the eastern side. Before that a ferry crossed the river here after a previous bridge was demolished in 1621. It is made from local red sandstone and has seven arches. At the time of its opening, it was the largest bridge

Smeaton's Bridge over the River Tay in Perth, dating from 1771, gets its name from the engineer who was in charge of its construction.

in Scotland. It was widened in 1869 by A. D. Stewart and there is a plaque on the bridge noting this. There are cantilevered walkways supported by iron brackets on the south side of the bridge. It is Category A listed.

Union Chain Bridge, Fishwick, Berwickshire/Horncliffe, Northumberland

The Union Chain Bridge, or Union Bridge, is a suspension bridge which crosses the River Tweed between Horncliffe, Northumberland, in England and Fishwick, Berwickshire, in Scotland. It lies 4 miles (6.4 km) upstream of Berwick-upon-Tweed. It was opened in 1820 and back then it was the longest wrought-iron suspension bridge in the world until the Menai Suspension Bridge was completed in 1826. It has a span of 449 feet (137 metres). It is a Category A listed building in Scotland, and a Grade I listed building in England. It is also an International Historic Civil Engineering Landmark. The bridge was designed by a Royal Navy officer, Captain Samuel Brown, who had joined the Royal Navy in 1795. When he saw the way that hemp ropes often failed on navy ships he experimented with wrought-iron chains as a stronger alternative. After successfully using these iron ropes on HMS *Penelope*, he built a small-scale suspension bridge to show that they could work on bridges too. He was then taken on as the engineer for the bridge, with work starting in

The Union Chain Bridge across the River Tweed between Scotland and England nearly closed in 2013 due to its dilapidated state. However, a campaign led to it being restored and it finally reopened in 2023. (Liam Tiernam)

August 1819. The bridge was officially opened on 26 July 1820 and tolls were charged until 1885. The bridge has a single span of 449 feet (137 metres) and is made from sandstone.

Over time the bridge has needed repairs and upgrades. In 2007 one of the bridge hangers failed, resulting in the bridge being closed until it could be repaired. Then in 2013 the bridge came close to being closed completely and was placed on Historic England's Heritage at Risk Register. In response, a campaign was started to restore the bridge, especially as its bicentenary was due to be celebrated in 2020. A restoration project was put in place and various councils, Historic England and the National Lottery Heritage Fund all pledged financial support. Work to restore the bridge began in October 2020 in spite of the Covid-19 pandemic. The bridge finally reopened on 17 April 2023.

White Cart Bridge, Renfrew, Renfrewshire

White Cart Bridge at Renfrew in Renfrewshire to the west of Glasgow carries the A8 over the White Cart Water, a tributary of the River Clyde. It is a Scherzer rolling lift bascule bridge, designed to open for river traffic going to the docks in Paisley. It opened in March 1923 and was built by Sir William Arrol and Co. It is the only rolling lift bridge still standing in Scotland and was last opened in August 2007 for a barge to pass under it. In the 1960s the bridge came under ownership of the local council and in December 1994 was declared a Category A listed building. In 2005 it underwent a £26 million refurbishment and the previous blue colour was replaced with a more distinctive maroon and cream colour scheme.

This unusual bridge known as White Cart Bridge is a Scherzer rolling lift bascule bridge which opened in 1923. (Allan Heron)

6

Wales

Chartist Bridge, Blackwood, Monmouthshire

The Chartist Bridge at Blackwood in Monmouthshire, South Wales, is a 745-foot-long (227 metre) cable-stayed bridge carrying the A4048 and B4521 across the valley of the River Sirhowy to the east of the town. It is supported in its centre by a 295-foot-tall (90 metre) A-frame pylon which supports the road deck. It connects Blackwood in the

The impressive Chartist Bridge at Blackwood was opened in 2005 and gets its name from the Chartist movement of the 1800s. (David Watkins)

west with Oakdale in the east. The bridge was constructed by Arup as part of the Sirhowy Enterprise Way road scheme and was opened on 23 December 2005. It is wide enough for three lanes of traffic, though the two flows are separated by white diagonal stripes. There is a footpath on the southern side of the bridge. As the bridge was built in an area where coal mines were common, the likelihood of subsidence was a major factor that affected its construction. This was taken into account by making sure that the bridge could 'breathe' if settlement did occur.

The bridge and the nearby Sirhowy Enterprise Way are operated and maintained by the DBFO Company (Design, Build, Finance and Operate) in accordance with the DBFO contract which lasts for a period of thirty years. The name of the bridge comes from the Chartist movement, a popular uprising in the nineteenth century with the aim of securing political rights for working-class men. Many of its supporters in South Wales came from the Blackwood area. In recognition of this there is a 26-foot-tall (8 metre) statue of a Chartist at the eastern end of the bridge, facing towards Newport, where the Chartists from Blackwood marched to. The statue is made of thousands of brass rings to signify strength in unity.

Chepstow Bridge, Chepstow, Monmouthshire/Tutshill, Gloucestershire

Chepstow Bridge links Chepstow in Wales with Tutshill in England across the River Wye. It is also known locally as the Old Wye Bridge or Town Bridge. It is made of cast iron and was opened on 24 July 1816, replacing several previous timber bridges on the same site. Originally, John Rennie, who designed Waterloo Bridge in London, was commissioned to design a new bridge, but his design was considered too costly. So, the firm of Hazeldine,

One of the prettiest bridges in this book, Chepstow Bridge dates from 1816 and was built to withstand the large tidal range of the River Wye beneath it. (Stephen Bunch)

Rastrick & Co. of Bridgnorth won the contract. John Rastrick was the designer and he altered Rennie's original designs and so reduced the cost by more than half.

As Chepstow Bridge crosses the River Wye with one of the highest tidal ranges in the world, it needed to be robust enough to stand the tidal pressure, and high enough to be above the highest tides. So, the cast-iron structure rests on four stone piers, each with their own cutwaters. At its centre the road deck is 13 feet (4 metres) above the highest tides and 57 feet (17.5 metres) above the lowest tides. The bridge has five arches, with the central span being 112 feet (34.14 metres) long, then the next two outward being 70 feet (21.34 metres) and the two spans nearest the banks being 34 feet (10.36 metres). The bridge is 20 feet (6 metres) wide and can only carry single-file traffic, which is controlled by traffic lights at either end. Until 1988, when a new bridge was built further downstream, it carried the A48 road linking Newport in Wales with Gloucester in England. It is Grade I listed and is the world's largest iron-arch road bridge dating from the first fifty years of iron and steel construction up to 1830.

Crickhowell Bridge, Crickhowell, Powys

Crickhowell Bridge carries the A4077 road over the River Usk at Crickhowell in Powys. It is believed to be the longest stone bridge in Wales at approximately 420 feet (128 metres) long, with a width of between 13 feet (4 metres) and 18 feet (5.5 metres). It is thought that the bridge first existed in the medieval period, as a wooden bridge, though the first official mention of it is 1538. It was rebuilt in stone in 1706, but after severe flooding a hundred years later, the bridge was rebuilt in parts, with the north-west side being widened in 1810 at a cost of £2,300. The engineer was Benjamin James. Then in 1828–30 the two largest

Crickhowell Bridge over the River Usk, near Abergavenny, is said to be the longest stone bridge in Wales. (Bethan Devonald)

arches on the northern side were combined into one larger arch. So, on the downstream side there are thirteen arches, but only twelve on the upstream side. It has V-shaped cutwaters on both sides and these have pedestrian refuges above them. The bridge had to be repaired in both 1928 and in 1979 due to the damage from the large number of road vehicles using the bridge. It is both a Grade I listed structure and a scheduled monument.

Devil's Bridge, Pontarfynach, Ceredigion

Devil's Bridge (Welsh: *Pontarfynach*) means 'The bridge on the Mynach'. It is situated by the village of the same name around 10 miles (16 km) east of Aberystwyth in West Wales.

It consists of three bridges on top of each other, each one being built many years apart. It carries the A4120 road over the Afon Mynach, which is a tributary of the River Rheidol. The name of the bridge comes from a tale where an old woman is said to have lost a cow which was on the other side of the river. She then did a deal with the Devil to build a bridge over the gap in exchange for the soul of the first living being to cross the bridge. The woman tricked the Devil by throwing a piece of bread across the bridge for her dog to retrieve, thus ensuring that the Devil did not get a human soul.

What is true is that there was a wooden bridge recorded as being here in 1188. This was replaced by the lowest bridge, a stone arch, believed to be medieval. Then in 1753

Devil's Bridge near Aberystwyth is the most unusual bridge in this book in that it consists of three bridges built together. (Theodore Lyth)

a second bridge was built over the lower bridge with another arch to hold the roadway. Finally, in 1901 the third iron bridge was added with the intention of keeping the roadway level. Seventy years later the bridge was strengthened and repaired. Today the bridge is a popular tourist attraction, being built across a narrow ravine with a steep drop of 300 feet (90 metres). A set of steps known as 'Jacob's Ladder' lead to the base and then onto the place where the River Mynach meets the River Rheidol. The terminus station of the Vale of Rheidol steam railway is also nearby.

Llangollen Bridge, Llangollen, Denbighshire

Llangollen Bridge crosses the River Dee in Llangollen, Denbighshire, and consists of four different-sized arches as well as a square opening across the railway line on its north side. It is constructed of coursed rubble, apart from the part which crosses the railway line, which is made from concrete. It is approximately 1,000 feet (305 metres) across and is 36 feet (11 metres) wide. It is Grade I listed and appears on the list of the Seven Wonders of Wales.

The actual date that the bridge was built is open to debate. It is thought that there has been a bridge here since at least 1282 when the nearby Valle-Crucis Abbey was built. However, there are two possible dates for the building of the current bridge. The first is approximately 1540, based on stonework within the bridge which states that it was rebuilt following the Dissolution of the Monasteries under Henry VIII.

Llangollen Bridge crosses the River Dee at Llangollen and is a four-arch stone bridge.

The second is 1656, based on the Rondie Reade stone which refers to payment made for construction work at the bridge. What is clear though is that the bridge was extended further than its four original arches to go over the new railway line in 1863. It was then widened in 1873 and again in 1968. It now carries two lanes of traffic, bordered by two pavements.

The famous Llangollen International Eisteddfod takes place every year not far from the north side of the bridge. In 2021 when the Eisteddfod was forced to go online due to the Covid-19 pandemic, Llangollen Bridge was decorated with a patchwork of colourful quilts. It was the work of the artist Luke Jerram, with the idea of representing all the corners of the globe.

Pont Fawr Bridge, Llanrwst, Conwy County Borough

Pont Fawr Bridge (or Big Bridge in English) crosses the River Conwy between Llanrwst and Trefiw in Conwy County Borough in North Wales and dates from 1636. It replaced a previous bridge on the same site which had been destroyed by heavy flooding. Some historians believe that the bridge was designed by Inigo Jones, the celebrated architect who introduced Italian Renaissance design into Britain in the 1600s. Although this assertion has not been proved either way, Jones was the Surveyor General under Charles I at this time. The bridge consists of three segmental arches joined by two piers and has a total span of 167 feet (51 metres). The central arch is 61 feet (18.5 metres) across, whilst the other two are 45 feet (12 metres) across. It was built using local gritstone mixed with slate rubble from nearby quarries. In 1675 and 1705 the bridge had to be rebuilt after further floods. It has two coats of arms on it, those of the Prince of Wales on one side with the initials *CP*, and those of the House of Stuart on the other side with the initials *CR*.

The stone bridge at Llanrwst which crosses the River Conwy is said to date from 1636. (Alan Greenhill)

Due to the steepness of the roadway across the bridge it is impossible to tell if traffic is coming the other way. Also, the roadway is only 13 feet (4 metres) wide, meaning that only one lane of traffic can use the bridge at any time. As there are currently no traffic lights to control the traffic, locals have nicknamed the bridge the 'Swearing Bridge' due to the frequent stand-offs between motorists who won't back down to oncoming traffic. It is both a Grade I listed building and a scheduled monument. It is also known as Llanrwst Bridge and carries the B5106 across it.

Waterloo Bridge, Betws-y-coed, Conwy County Borough

Waterloo Bridge (Welsh: *Pont Waterloo*) crosses over the River Conwy at Betws-y-coed in Conwy County Borough and dates from 1815–16. It was originally called the Llynnon Bridge but was renamed after the Battle of Waterloo which took place in 1815. The words 'This arch was constructed in the same year the Battle of Waterloo was fought' are written across the arch under the roadway. Although the bridge was constructed mostly in 1815 and has the date 1815 in its centre, it was not finished and officially opened until 1816. It also has decorative castings of the four national emblems in its spandrels: the leek for Wales, the rose for England, the thistle for Scotland and the shamrock for Ireland.

The bridge was built to carry Thomas Telford's London to Holyhead road over the River Conwy; this would later become the A5. It was a replacement for another bridge, Pont-yr-Afanc, further to the south. This bridge had an approach which was too steep for

Waterloo Bridge was designed by Thomas Telford and carries the A5 London to Holyhead road over the River Conwy. (Les Williams)

horse-drawn carriages to move along easily and so Telford designed Waterloo Bridge with a more level approach. It has a span of 105 feet (32 metres) and is made of cast iron, apart from its stone bastions. The bridge was upgraded in both 1923 and 1978 with concrete added to the substructure to strengthen it. It is a Grade I listed building.

Wye Bridge, Monmouth, Monmouthshire

Wye Bridge in Monmouth, South Wales, is a Grade II listed building which was rebuilt in stone in 1617, having previously been an earlier wooden medieval bridge. It is situated to the east of the town and until the A40 bypassed the town was the main bridge across the River Wye in Monmouth. It now carries the A466 and meets the A40 at its western side. This arched stone bridge was built using red and buff sandstone ashlar and has five arches. The bridge was widened between 1878 and 1880. It is 233 feet (71 metres) in length and is a Grade II listed building. It has two pillboxes on it which were built as defences in the Second World War, as part of Western Command Stop Line No 27. These are now scheduled monuments.

The final bridge in this book is Wye Bridge at Monmouth dating from 1617. (Bethan Devonald)

Other Information

Websites

bridgemeister.com
cliftonbridge.org.uk
historicbridges.org
nationaltransporttrust.org.uk
sabre-roads.org.uk

Further Reading

Chatterton, Mark, *Britain's Coastal Road Bridges*
Chatterton, Mark, *British Road Bridges – An Introduction*
Cruickshank, Dan, *Dan Cruickshank's Bridges: Historic Designs that Changed the World*
Denison, Edward and Stewart, Ian, *How to Read Bridges*
Fisher, Stuart, *Rivers of Britain*
McFetrich, David, *An Encyclopaedia of British Bridges*
Rogers, Jospeh, *Britain's Greatest Bridges*

Acknowledgements

I would like to thank the following people and organisations for their help in providing support, photographs and information for use in this book:

Cheryl Billington, Stephen Bunch, Richard Chatterton, Nick Clacy, Lewis Clarke, Bobby Clegg, Bethan Devonald, Alan Greenhill, Paul Harrup, Allan Heron, Andrew Lamyman, Helene Lyth, Theodore Lyth, Peter McDermott, Dick Meads, Alan Morris, Ian Park, Mike Pollock, Liam Tiernam, David Watkins, Ian Watts, John Webb, Les Williams, Pete Wing, Richard Woolham.

Bridge of the Day Facebook group

Janice Boakes at Berkhamsted Local History & Museum Society

Sabre-roads.org.uk

All photographs are by the author unless otherwise stated.

Also by Mark Chatterton

British Motorways – An Introduction
British Road Bridges – An Introduction
British Road Tunnels – An Introduction
Britain's Coastal Road Bridges
Britain's Motorways
Britain's Road Tunnels